Building a Successful Construction Company

The Practical Guide

Paul Netscher

Published by Panet Publications
PO Box 2119, Subiaco, 6904, Australia

www.pn-projectmanagement.com

ISBN: 978-1500680008

Available from Amazon.com and other retail outlets

Legal Notices

It should be noted that construction projects are varied, use different contracts, abide by different restrictions, regulations, codes and laws, which vary between countries, states, districts and cities. Furthermore various industries have their own distinct guidelines, acts and specific protocols which the contractor must comply with. To complicate matters further these laws, acts and restrictions are continually evolving and changing. Even terminologies vary between counties, industries and contracts and may not be the same as those included in this publication. It's therefore important that readers use the information in this publication, taking cognisance of the particular rules that apply to their projects and area in which their company operates.

Each project has its own sets of challenges and no single book can cover all the steps and processes in every project. This publication covers a broad range of projects without being specific to a particular field of work. Some of the author's personal opinions may not be pertinent to certain projects, clients, circumstances or companies. Readers should undertake further research and reading on the topics particularly relevant to them, even requesting expert advice when required.

Therefore, the author, publisher and distributor assume no responsibility or liability for any loss or damage, of any kind, arising from the purchaser or reader using the information or advice contained herein.

The examples used in the book should not be seen as a criticism of people or companies, but, should rather be viewed as cases from which we can all learn. After all we've all made mistakes.

Any perceived slights are unintentional.

Cover layout by Clark Kenyon, www.camppope.com
Photographs: Cover - © iStock, Title Page – Ian Weir, Preface – Paul Netscher, Acknowledgements - © iStock

Preface

Managing a company isn't easy. Managing a construction project is hard. Put the two together and manage a construction company – you have a challenge. By nature construction projects are short term, many are high risk and most are low reward. In addition the construction industry is always changing – going from a few years when work is bountiful and then rapidly changing to years when work is in short supply. One would think the good years are easy! Unfortunately they often aren't. Yes, it may be easy to find work, but suddenly skills, materials, subcontractors and even equipment are no longer available, or if they are they've quickly become more expensive reducing the good profits the company was hoping to make. To grow and be successful, managers of construction companies need to be skilled and astute to guide their company on this rollercoaster ride. They need to be able to adjust their thinking and way of operating to suit the changing circumstances. Even one wrong move; undertaking the wrong project, submitting a quotation that's too low, misreading a contract, underestimating the schedule, making a major mistake on a project, or miscalculating the cash flow can quickly destroy the company, ending many years of hard work.

After 28 years in the industry I've seen some companies grow and be successful year after year, while others have barely managed to keep afloat. A number have even enjoyed a meteoric rise and a few years of success before collapsing and going out of business. In this book I've tried to pass on some of my experiences. This book isn't going to show you how to run a company (there are many other books that do that), rather, it's aimed at giving practical tips on how to manage and grow a construction company and to maximize profits.

Acknowledgements

Thank you to all the people who have worked with me in the last 28 years. Many of you have in some way contributed to this book.

Thanks to my family for supporting me through this writing process.

Thank you to Sally for editing the book.

A special thanks to Tim for also editing the book and providing many invaluable comments.

Contents at a Glance

Contents

Introduction

Many people think there's lots of easy money to be made in construction. They may be working for a construction company and see owners and senior managers driving expensive cars, or they hear of companies doing well and declaring big profits. It seems to be a fast way to make money. So, many good tradesmen give up their jobs, invest their life savings and start their own construction business. Unfortunately, the truth is that for every rich and successful person in the construction industry there are probably at least ten others, business owners and managers, who aren't as successful, working long hours and taking home an average salary. Furthermore, there are probably several business owners who've lost their companies and are once more working for a boss, sometimes even in another industry. Nearly all of these people are skilled and knowledgeable and I'm sure every one of them worked hard. Many of them probably even completed successful projects, and yet, their company wasn't successful and eventually collapsed.

People who have owned or managed a construction company that has fared poorly often ascribe the poor performance to bad luck. Often it's just one bad contract that has wrecked their company. It may just have been one client that didn't pay them, or it could have been one project that was affected by unseasonably bad weather. I've worked on projects that have lost money, where we could attribute the loss to bad luck. Maybe there was bad luck! Maybe it did rain and possibly we did have equipment breakdowns. But, to be honest, if we analyse the reasons for the poor performance, we probably could have avoided the loss, or at least reduced it, if we had done things differently and managed the project better.

To be successful in construction means more than just completing numerous projects. It requires that the projects must be completed on time, to the required quality standards, with no safety incidents, and importantly, make a profit. But more significantly, the company must be paid for the work they've done and be cash positive. Cash flow is one of the biggest challenges facing construction companies. Just because a project is profitable doesn't mean the company has money in their bank account. Furthermore, the company needs to build a good reputation, establishing relationships with clients, so they can obtain further work.

Construction isn't an easy business with clients that are sometimes difficult, demanding, inexperienced, have unreasonable expectations and even in some cases may be simply dishonest. Clients also set unreasonable schedules, have their own cost pressures (forcing construction companies to provide unsustainable savings), and appoint substandard design teams that provide poor quality drawings and information, often late, and aren't responsive to the construction company's needs and queries. Furthermore, contractors often face a shortage of materials, additional red-tape, tougher safety and environmental legislation, a society that has become litigious, added competition, more complicated and complex projects and in some cases even political interference.

Every project and client is different and contract conditions are often varied. Therefore, there isn't one set recipe that will work on every project, forcing construction companies to continually adapt and modify their processes. Moreover, contractors still face the same challenges of adverse weather conditions, price increases and a shortage of skills.

However, if the truth be told, there are very few well run construction companies. Many fail, not because of the outside pressures, but rather because of their own inadequacies. Some reasons for their failure are; poor tendering, lack of cash or poorly managed cash flow, over-extending themselves where they have insufficient finance, resources or skills, not getting paid for work, poor financial controls, a lack of leadership and a shortage of work.

So what makes a construction company financially successful? Well firstly managers must find a project to price – and not just any project, but the right one. They need to accurately tender or price this project and make sure they win the work without giving all their potential profit away. Once the company has been awarded the project they must successfully do the work, avoiding mistakes and ensuring they're paid for all completed work. To achieve this they need to be financially and contractually astute and employ good people. At the same time managers need to be enhancing the company's reputation, managing the overall company, and growing the business.

There isn't a magic formula for this, but, there are some common sense rules to assist in achieving these outcomes which I've outlined in the following chapters. There's no coincidence that the two longest chapters are on tendering and reducing costs. If you're awarded a project at the correct price, ensuring the conditions aren't unreasonable and the schedule is achievable, and then carry out the work efficiently with minimal costs, the company is half-way towards becoming successful.

Being successful is not so much about being lucky, but rather about creating your own luck. Obviously it also demands hard work and a certain amount of technical knowledge and practical skills.

Managing and running a construction company requires knowledge of many particular laws such as; company registrations and compliance, labour legislations, legal and contractual knowledge, safety and environmental legislation and tax laws. This book doesn't go into detail on these since they vary between countries and change with time. It's however essential that companies comply with the prevailing legislation, so it's important that managers have a broad understanding of the requirements and seek expert advice and help.

I was fortunate and was employed for twenty years by a very successful construction company. The company expanded and overtook most of its larger competitors. As the company grew I was able to grow and develop, benefiting from the success of the company in monetary terms, in personal growth and job satisfaction.

Chapter 1 – The Right Project

Many construction companies are desperate to find work so they're willing to accept any project, at any price. I've been in a similar situation before. However the risk of accepting a project at any price is that the company wins the project, works hard on it for several weeks, or months, only to lose money. How much better would it have been for everyone to sit at home, neither making money nor losing it? By not working on the project maybe the company would have had time to look for other more suitable projects? Maybe, by taking on the project the company didn't have resources available to undertake a more suitable project when it presented itself?

The risk of taking on the wrong project not only impacts directly on the profitability of that project, but it impacts on other projects and the company as a whole. When a project is in trouble management spends more time on it than they normally would, trying to rectify and solve the problems. This time could be better spent on maximising profits on other more successful projects. In fact, the absence of management on the other projects sometimes results in them also turning bad.

Working on a project should not be about being employed. The only reasons you want to be doing a project is so the company can make money, or occasionally so it can win further work which will make money. Frequently I see contractors win projects just because they're desperate for work, or, sometimes contractors take on large and prestigious projects, believing it will enhance their standing with the public and their shareholders. Unfortunately, many of these projects end up costing the company large amounts of money.

So what is the right project? It's one which the contractor can:

- complete on time
- produce good quality work which satisfies the client's specifications and requirements and is a good advertisement for the contractor
- complete with no safety incidents, without harming anyone or the environment
- make a profit
- receive payment on time
- be cash-positive, or at least cash-neutral
- utilise their available resources
- if possible, obtain further work from the same client or win work from other clients in the vicinity

Know which projects are profitable

If a contractor understands which contracts are their most profitable, it makes sense to place more emphasis on them and pursue that particular field or client. This sounds obvious, but unfortunately some contractors don't always know how much profit each project has made, nor have they analysed the reasons why some projects are more profitable than others. This results in contractors submitting tenders for projects which won't be profitable, even tendering for projects where they consistently lose money.

Case study:

We constructed two concrete cooling structures and lost a considerable amount of money on both projects. At first glance the projects appeared simple and were priced as such, but in fact each time the construction turned out to be more complex and difficult than anticipated. The first time we lost money we thought it was because we ran the project badly. After we lost money the second time we realised it was best to avoid these projects.

I later talked to a number of other contractors and none of them seemed to have made money on these structures.

After turning down the invitation to tender for several more of these projects I eventually couldn't resist the temptation to price one and try and prove we could actually tender it correctly and build it efficiently without losing money. Fortunately we were able to win the project at a relatively high profit margin of 20%. Needless to say, we made only 12% profit on the project, meaning we lost 8% of our tendered profit.

The above just proves that some projects are best left for other construction companies to build.

If you lost money on a project, before you tender on a similar one, you need to understand the reasons for the loss and ensure the same problems won't arise next time around.

Working for the best clients

Picking the right client and working for them is probably one of the essential keys to being a successful contractor. So what makes a good client?

- Most importantly the client must pay for the work done. Not only must the client pay the contractor, but they must pay the contractor on time and in full. This means they must be financially sound with money in the bank, a positive cash flow, and have no risky ventures or businesses.
  So how do you enquire about the financial wellbeing of a client? You can ask for a copy of their bank statements, but many clients are unwilling to divulge these. Anyway having money in a bank doesn't mean the client will pay the contractor on time, or even at all. So a contractor should do a bit more research. By reading newspapers one could hear about a client's financial problems. In addition, talk to subcontractors, suppliers and other contractors and you will get a sense of whether the client is a good payer or not.

- It's also useful to work for a client that may have other work, either on the same project or elsewhere. Many of the projects I was involved with went onto a second phase, one even went to five phases, and we often remained on a site for several years. We worked for a large petrochemical company and were often able to work on several projects simultaneously while established on their facility.
  For a time our competitors made sure they kept us away from certain mining clients. When we eventually won a contract with these clients we were able to undertake several more large projects which were very profitable and ultimately made up a significant portion of our overall profits. We quickly discovered why our competitors had been so protective of these clients.
- Some clients are more organised than others and employ strong professional teams. The more organised the client, the more likely it will be that drawings and information will be issued in accordance with the schedule, which will enable the project to be completed on time, within budget and with minimal fuss. Although late information and delays may give the contractor an opportunity to submit claims and variations, earning the contractor additional money, the delays are often counterproductive, tying up staff unnecessarily, preventing them from moving to the next project and causing them to become frustrated. Obviously, in a tight market, with little work and no project for staff to move to, it can be advantageous to be delayed on a project and paid by the client for this time. However, in most cases it is far better to complete the project in the allotted time (or faster), and move onto the next one.
- Some clients can be particularly fussy and demanding especially if they don't understand the construction processes properly. At times these clients will argue about every little detail and every cent. I'm not advocating that clients shouldn't insist on quality and value, but rather that some clients go beyond these requirements, becoming quite petty, frustrating the contractor's staff and tying them up in time consuming arguments. I've known some contractors that refuse to work for certain difficult clients, or deliberately add a premium in the tender price to allow for the trouble of working with them.

Case study:

We undertook a road project for one client who used to send us letters daily, some even late at night. These letters usually required a response from our Project Manager, but were often about small details that many clients would ignore. In fact, often the facts and logic were incorrect. All of this tied up our Project Manager and prevented him from working on other projects. In addition, the client's constant interference in the work methods slowed us down and if we hadn't been contractually astute the client would have deducted money from us which they weren't entitled. All of this added additional costs which weren't budgeted for and we lost money on the project.

We later heard that another contractor who had previously worked for the client added a 15% premium to their tender price to compensate for the additional costs they knew would likely be incurred on the project.

The strength of the client's team

The client usually appoints a team of designers, sometimes architects, and occasionally a managing contractor. Some of these companies can be very professional and often assist the contractor in delivering the project on time. However, I've found some to be inexperienced, even obstructive in the way they administer the project, and on occasion unresponsive to the contractor's requests for information, drawings and access.

The contractor depends on the designers to produce quality drawings and reliable information on time. Inadequate drawings hinder the project progress which could negatively affect the contractor's costs.

Projects run by a poor team lead to the contractor requiring a larger administrative team which adds to the contractor's costs. Also, these projects often result in delays, claims, variations and disputes which further add to the contractor's administration costs, negatively impact their cash flow, and even affect their reputation. Projects with these teams on board are therefore best avoided.

I've worked with many good teams and clients. The projects have generally been well organised, drawings have been of a high standard and issued on time. They've been responsive to our questions and have been fair in their assessments of our variations. It's not to say there weren't mistakes or problems, but in general things were done professionally. The end result was always that we had a successful project, finished on time, with minimal fuss, with a happy client, but, more importantly we always made money – usually significantly more than anticipated at tender stage.

The field of construction

What field or type of project should you be tendering on? I had one particular business owner who was prepared to tender on anything, even if we had absolutely no experience in that field of work. Clearly, to prepare a proper tender you need some knowledge of the field to enable an accurate price to be developed. The client would also typically only consider awarding a project to a contractor who has the relevant experience in the type of work to be undertaken. This is not to say that contractors shouldn't be looking at alternative fields in which to work, but, these should only be considered after considering all the risks and opportunities.

Often branching out into a new field requires the contractor to employ staff with the required knowledge, even purchasing specialist equipment. Therefore there should be some certainty that after completing the project these resources can be employed on other similar projects.

Occasionally, opportunities arise on an existing project, or with an existing client, to undertake a project in a different field. If there are benefits in undertaking the project consideration should be given to either subcontracting the specialist work, or forming a joint venture with another company that has the required expertise.

I wouldn't advocate tendering for projects in new fields without first ensuring there's sufficient knowledge available, either within the company or from outside experts, that can assist with the tendering and construction.

Case study:

One of our competitors decided to start a division to undertake slip-form construction. Since there were many opportunities for slip-forming work, with very few competent contractors, the idea was a good one. However they didn't employ the correct personnel for their first two projects and these weren't executed well. This unfortunately cost them money and also negatively affected their reputation.

It should be remembered that any work done poorly, even if it is done by a new division, can adversely impact the entire company's reputation.

Opportunities for further work

A good project to work on is one where there's potential to be awarded further work. It's often advantageous to be one of the first contractors on a large project. Some contractors even tender for a project at a reduced or even zero profit to ensure they win it in the hope they'll be awarded further work where they can recover their profit. This can be dangerous because they might not get further work on the project.

Some projects might be the first part of the client's facility and they have indicated that they will be constructing further phases. Unfortunately these phases often don't follow immediately after each as a continuous stream of work.

It's still worth doing homework, understanding the bigger picture and the potential for more work from the same client, another client in the same facility, or even within a particular area. For instance, if your company specialises in home renovations, then a suburb that's undergoing redevelopment with many owners doing renovations may be a good area to be working in. If prospective clients see the company's name on sign boards, vehicles and equipment there's a good chance of being asked to quote for other projects in the area. However, working in an established suburb, where renovations and developments are few and far between, won't necessarily generate the same exposure for the contractor.

The project's location

A project adjacent to, or in the vicinity of the company's Head Office may be desirable. Not only are these projects convenient to manage and support, but it wouldn't be good to have a competitor constructing a project so close to the office, particularly when potential clients visit the office.

Some project locations are prominent, offering an opportunity to advertise the company. The successful completion of the project can result in good publicity giving the company an opportunity to showcase its abilities.

On the other hand, some projects may be difficult, disrupt business and traffic and result in adverse publicity, particularly if they are in prominent locations.

The project's proximity to existing projects

One business I worked for tried to price anything and everything, with some projects being thousands of kilometres from where we were operating. We weren't running the projects close to home very well so it was likely that these distant projects would be run more poorly.

Don't underestimate how much travelling eats into management time. When projects are located close together Project Managers and Project Directors can often easily look after more than one project, and they can certainly look after more projects than when they are widely spread over large distances.

Projects situated close together may also have other efficiencies such as sharing resources like mechanics, fitters, service trucks, fuel bowsers, Site Administrators, Contract Administrators, transport, office space and accommodation.

Remote sites

Many companies are good at working in remote locations. In fact some specialise in working in remote areas, though, most don't operate successfully in these locations.

Projects in remote locations have their own challenges:

- These projects require personnel who are prepared to work in these locations, yet, most people are unwilling to do so, and many companies just don't have the right personnel for such projects. These companies either recruit people specifically for the project, which isn't a good idea since they are relying on someone who doesn't necessarily understand the workings and culture of the company, or, they force their existing staff to relocate, leading to them being unhappy and either not focusing on the project or resigning. Why lose a good person just because you have a project in a remote area?
- It also takes a special kind of person to work in a remote area – someone who's resourceful, independent and who can get on with the project with little or no assistance.
- Working in a remote area demands certain logistical skills because personnel, equipment and materials must be transported over long distances to the project. The projects are often difficult to reach and may present difficulties in servicing and repairing equipment, supplying spare parts and even getting paperwork back to Head Office.
- Remote projects must be tendered correctly, and the Estimator must be aware of additional salary costs and allowances. Since plant and equipment may not be readily available it will be expensive to move equipment onto and off the project. Equipment may sit idle on the project between tasks because it's uneconomical to remove it and return it later when it's required. Maintenance and repairs will be more costly, and may take longer to implement.
- Management visits won't occur as frequently as for projects closer to Head Office, meaning problems could go undetected or unresolved. Many of my projects were several hundred kilometres from our office

and I required a full day to travel and visit them. I envied Project Directors who had all their projects in the city and could easily visit a project whenever they had a few spare hours in the day.

- If the area is remote there's often not much chance of further work in the vicinity. Of course, if there's a possibility of other work in the area that can be done at the same time, or immediately following the project, the company could be well placed to be awarded the work and some transport costs may be shared with the new project which would make them both more profitable.

The resources required for the project

Consideration must be given to the types and quantities of resources that will be required to carry out the project.

- Some projects can be fairly difficult, or spread over a large area, and require lots of management and supervision for a relatively small revenue and profit. Other projects with a similar profit and revenue may require less supervision and management, so it's obviously better to focus on these because the company is able to do a larger turn-over (hence making more profit) using the same number of people.
- Of course some projects are better suited to certain staff, so it's advisable to tender on projects that match the abilities of staff requiring work. For instance if the company is looking for work for its Building Supervisors and doesn't have spare Concrete Supervisors they should be tendering for building rather than civil projects.
- If a company owns a significant amount of equipment and most of it is unutilised then it's useful to win a project (even at a lower margin), that will utilise the equipment and earn revenue for it.

Unfortunately some companies tender for and win projects for which they don't have enough suitable staff. This means they have to employ new people for the project. Because the recruiting is usually done in a hurry some of the new staff may not be the best or the most suitable and are just bodies to fill a position. The new engagements are normally not familiar with the company's procedures, safety and quality expectations or behavioural codes. This often leads to problems which may cost the company money as well as its reputation.

Keep it simple

Often the simplest and least attractive projects are the more profitable ones.

Case study:

For many years our company was unable to secure a project with a particular client who was usually constructing a number of new projects at any one time. One day we got our opportunity. A major competitor had constructed a new process plant for the client. The main work was complete, but as with many large projects there were lots of bits and pieces to be finished off. The other contractor wasn't interested in doing these and only wanted to do larger projects. The managing contractor approached us to undertake the remaining works which we accepted, since we were only too happy to finally work for the client.

Not only was this work profitable for us, but after completing the project the client gave us opportunities to undertake further projects, leading to us eventually becoming their preferred contractor.

We always said that when a project won an award, or had fantastic photographs that appeared on calendars, then it had almost certainly lost money. Obviously this is a generalisation and isn't always true. However, it's often the case that projects that look fantastic are the more difficult and costly ones to construct and are usually more risky. Many of my most profitable projects aren't visible to the general public.

One of our major competitors always chased mega projects which were good for headlines, but they seldom earned the major profits they expected.

The project's risks

It's important the contractor understands the risks involved with a contract since some projects can be very risky. These risks could be related to a weather event, (do you really want to build a bridge across a river during the rainy season?). They may be related to possible industrial relations problems, or there might be risks of damaging public or private property, or injuring someone.

Case study:

One of our good clients was very persistent that we undertake a small very difficult project. We had previously completed several large projects for them and at that time were busy on a couple of projects worth several million dollars. This particular project was worth only a hundred thousand dollars and was in the middle of a petrochemical plant. I mean in the middle! It was directly under and around major operational equipment and we were expected to excavate, locate and remove literally hundreds of existing underground cables. The project was high risk, and if we damaged operating equipment or cables we could have shut down portions of the plant which would have resulted in bad publicity for our company and probably have prevented us getting further work from the client. In addition, we may have been liable for the costs of the repairs. Obviously the profit from a one hundred thousand dollar project couldn't make up for these risks so this was definitely a project to be avoided.

Unfortunately, in this case the client was really persistent and we were eventually forced to accept the project. Nonetheless, we bought additional insurance in case we damaged any of the existing facilities, made sure the client changed some of the conditions in the contract document (reducing our risk exposure), and included sufficient supervisors and management in our price to lessen the chance of problems occurring. We ended up successfully completing the project without mishap and made a good profit.

From the above I'm not saying that you shouldn't tender for a high risk project, but it's important to consider and understand the consequences of the risks. You don't want to destroy the company because a big risky project was undertaken and the potential risk events actually happened.

When tendering for projects with high risks there are various strategies that can be employed to mitigate these risks:

- Some risks can be reduced by negotiating better contractual terms with the client and transferring risks to them.
- Some risks can be partially allowed for in the pricing – for instance if we expected to lose time due to inclement weather we often priced to allow for a portion of the expected lost time. Obviously if there was no inclement weather we wouldn't have spent the contingency and would have made more profit. If we experienced more inclement weather than allowed for we would have spent the entire contingency and used some of our profit – but at least by allowing for some inclement weather we had reduced the risk and potential loss.
- In addition it may be possible to purchase insurance to cover some risks.
- Sometimes one can add additional profit to the tender, making it worth taking on the risks.

Of course all construction projects have some risks and it's a matter of ensuring that the risks are understood and allowed for.

Some risks to consider are:

- difficult ground conditions such as having to excavate in rock
- existing services which can be expensive to locate, result in delays when they are moved, and if damaged can result in large repair costs and harm the company's reputation
- adverse weather conditions, such as rain, wind and severe temperatures
- industrial relations problems
- unknown or untried technology
- a short or difficult schedule
- complex and difficult structures
- working in and around existing operating facilities

The right size

Just like the children's story of Goldilocks, where some items weren't right (either too big or too small) and others were just right, you don't want to be working on projects that are too big or, too small. You want to work on projects that are the right size for the company. I must emphasise *'right for the company'.* Consequently it's important to understand what the best size of project is. A project that is suitable for a small company may be too small for a larger one. As the company grows, so will the size of their projects change. This may even affect the clients the company works for, and even the market or region in which they operate. Many contractors grow in size but are reluctant to abandon their existing clients. I'm not advocating dropping all existing clients, but occasionally, companies grow out of clients, and the company has to move on, politely declining to price projects that may now be too small or unprofitable.

Many contractors get into trouble when they win projects that are too big since:

- They have insufficient personnel, causing them to hire new staff who could be inexperienced and don't understand the company's values. Alternatively, the company puts too few staff on the project, resulting in poor management and supervision.
- They don't have suitable personnel with large project experience.
- The company doesn't have the right experience and knowledge to run a major project.
- Large projects place bigger demands on the company's cash flow.
- A large project may require all of the company's resources which means the company is unable to work on other projects, effectively meaning they have all their 'eggs in one basket', which could be disastrous for the company should the project go badly.
- If their resources are all tied up on one project they are unable to do projects for other clients, some of whom may be good ones, which causes these clients to use another contractor which may well result in the contractor losing them forever.
- When a large project ends there's often a sudden release of personnel and equipment and replacement work has to be found for all of them in a short period of time (ideally you want a number of projects starting and ending at different times – each with a value not exceeding fifty per cent of the company's annual turn-over).
- Large projects require bigger bonds and sureties which could take up all the company's banking facilities, preventing them from taking on other projects. It should be noted that many projects require these bonds and sureties to be in place until the end of the warranty period which could be a year or more after the project completion.

Just as a project can be too large for a company, so to, are there projects that are too small; either in duration or value. Smaller projects:

- Often require the same management and supervisory staff as a large one, so, it's possible for the same number of staff to look after and generate more revenue per month with larger projects than with a number of smaller projects.
- Are normally of a shorter duration than larger projects and are therefore less efficient because every time personnel and equipment are transferred between projects there are inefficiencies and lost time due to travelling between the projects and preparing for the new one. In addition one project seldom follows on exactly from the previous one. Consequently there's often a break of a few days, weeks or months, when personnel and equipment are idle and not earning revenue.
- Take more Head Office administration time than one large project does because:
    - payroll staff must administer frequent transfers of personnel between smaller projects

- creditor administrators must deal with many smaller purchases compared with a few larger purchases for the big project
- debtors staff must process more invoices to the many different clients compared with only a few invoices for the larger projects
- the estimating department must submit and win more tenders

It is however useful at times to have a mix of different sized projects since smaller projects allow junior Project Managers to manage their own projects, while larger projects employ a number of junior engineers who are able to learn from the more senior staff on the project.

Niche markets

Sometimes contractors have a particular niche field or client that they supply and service. It's generally in the contractor's interest to keep other contractors out of this niche so they must ensure their tender is competitive enough to do this.

In other cases, if the contractor has the experience, equipment and personnel suitable for the project then their costs may be lower than other tenderers who don't have these advantages, which will enable the contractor to apply a larger profit margin.

Contractors must be careful not to become trapped in a particular niche because often these niches can come to an end when other contractors enter the market, or when work dries up. If the contractor doesn't have other clients and fields to work in they suddenly find they have no work, leading to them going out of business.

The moral of the story is to develop and chase a niche market where practical, but always to be involved with other opportunities that can replace the revenue from this market should the work come to an end.

Fad or potential for a new market

Case study:

Several years ago many of the contractors working in South Africa were unable to find sufficient work in the country so they tendered for, and won, projects in other African countries with varying degrees of success. Operating conditions weren't easy, logistics were often a problem, some clients didn't pay, employees were unhappy working far from home, the legal situation was different and there were additional taxes which weren't always factored into the price. Unfortunately the success stories were outnumbered by failed projects. Many of the medium sized companies saw the large companies moving into Africa and decided to do the same thing, bringing more accounts of failure.

The moral of the story is that there was potential to make large profits in other African countries but it did depend on the country, the client and how well the tender was done. Just because another company has been successful doesn't mean your company will be as successful, and nor does it mean that when you move into that market the profits will be as easily made.

It's important to research new markets and opportunities to ensure you understand all the risks and that the market is going to be sustainable in the future. You don't want to be setting up an office in another country, state or region just for one or two contracts, nor do you want to purchase new specialist equipment for a new market only to find that the market dries up and you're left with equipment you cannot use.

However, if you've done your research properly and believe that the financial model works, that you can minimise the risks and the market is sustainable, then by all means tender for the project. Maybe you will find a profitable new market.

Understanding the current market conditions

Different projects can be more suitable than others depending on the prevailing market or economic conditions. When there's little work available it's not always possible to be selective about which projects to tender for.

The construction market changes and moves through cycles of a relative abundance of work to a scarcity very quickly. It's important to understand and anticipate these cycles. I've regularly seen contractors take on a difficult project at a low profit margin just as the market enters a boom period. Their resources are then tied up on the project with a low margin, while other contractors are free to tender for more profitable projects.

The opposite is also true, and it's important to secure a long duration project at the end of a boom cycle at a good profit, just before the tender profit margins start dropping. This project can at least get the company part way through the difficult period.

Profit margins must be adjusted appropriately and timeously to take into account these changes in market conditions. I've frequently seen contractors at the end of a boom period not adjust their margins downwards fast enough in anticipation of the downturn, and every time they do adjust the margins down they aren't reduced low enough to secure the project, forcing them to go even lower on their next tender.

Sometimes if you're aware your competitors are busy on other projects, or aren't keen on tendering for a project, it's possible to win the project with a larger profit margin than would normally be applied.

Bid (or tender) procedures

How the client calls for bids should influence the decision whether to price a project or not.

- Many clients put projects out to open tender which means there are no restrictions and any contractor can submit a tender. Government projects are usually put out to tender in this way. Obviously the client may receive dozens of tenders for the project and I've seen clients receive up to fifty tenders. The competition is fierce and the chance of winning the contract is low, particularly since there's often at least one contractor who is desperate and submits a low or even stupid, tender. With an open tender there are also often inexperienced smaller contractors who don't understand the project correctly, causing them to submit an incorrect price. Where possible I'd avoid submitting tenders for these projects.

- Another way clients call for tenders is anyone can tender for the project providing they meet certain criteria stipulated in the tender documents. Some of these conditions can be quite simple such as the contractor must be registered in the state or country. Some can be more restrictive and many contractors are unable to meet them. These may include a requirement for a tender bond which some contractors may not be able to provide, the contractor might have to demonstrate they have constructed similar projects, achieved a minimum annual turnover, or there may be a specified minimum local content or local contractor participation target. The number of contractors eligible to tender for the project depends on the number and types of conditions. If a contractor cannot meet the conditions of eligibility then it's usually a waste of time to submit a tender, although, some clients don't always adjudicate tenders in a fair and honest manner and could be swayed by a low price and award the project to a contractor who doesn't meet the tender criteria.
- Some clients call for contractors to prequalify to tender, and only contractors who successfully qualify are invited to bid for the project. Unfortunately some clients can prequalify a dozen or more contractors to submit tenders which can still mean that the competition to win the contract can be fierce.
- Most private clients invite only a select number of contractors, which is normally around six, but can be as few as three or as many as twelve, to tender. The competition is obviously much reduced and it's definitely worth tendering for these projects. The trick is to become part of the group the client invites to tender. More on this subject in the next chapter. An important item to bear in mind is that if a client invites a contractor to tender there is almost an obligation, and an expectation, for the contractor to submit a tender and if they don't submit one some clients can be unforgiving and possibly won't invite the contractor to tender for their next project.
- Of course the best method of winning a project is when a client negotiates the contract with the contractor. I've managed to arrange this on a number of occasions. Not only is it possible to achieve a slightly higher margin, but more importantly it almost definitely removes the opposition from the tender process thus ensuring the contractor can secure the project. During the negotiation process it may be possible for the contractor to offer up ideas for improving the constructability of portions of the project, thereby providing cost savings for the client and maybe even increasing the profits for the contractor. Having said this though, as a word of caution, I have seen contractors lose contracts they were negotiating because they were either too greedy and wanted the contract at a price the client wasn't prepared to pay, or because they were demanding terms and conditions unacceptable to the client.

The contract document

Clients use different forms of contract documents. Many of these are simple variations of standard conditions used throughout the industry. Most offer protection to both the client and the contractor. Some clients, however, use their own form of contract document which can sometimes be heavily biased, transferring all the risk to the contractor and offering them little protection. Either avoid tendering for these projects or ensure the risks are removed, qualified out, or priced in the tender submission.

Payment conditions

The cash flow on a project is important to the success of the company. Some clients have payment periods that are short – possibly even a week or two weeks after the submission of an invoice. These projects are far more attractive than those with terms of thirty-five days, fifty-six days or longer. Many smaller contractors have to pay suppliers upfront and most contractors have to pay suppliers within thirty days. If the contractor is paid later than this they are in effect financing the project. Many building contractors make more money from managing their cash flow properly than from actually doing the work, since they're often paid at seven or fourteen days and only pay their suppliers and subcontractors after thirty days.

The difference between projects paying after fourteen days compared to a project paying on fifty-six days could be worth ½% of the project value in interest earned, which might not seem much, but if the project profit is 5% this increases the profit by 10%.

But that's just on the payment period. Now what about some of the payment conditions? Clients often hold between 5% and 10% retention with half released at the end of the project and half released at the end of the warranty period which is normally twelve months.

A project with a duration of one year and a warranty period of one year, with a 10% retention held to the end of the project and 5% held over the warranty period could add nearly another ½% of interest costs to the project compared to one with no retention.

Moreover consider when and what the client pays. Many clients don't pay for materials and equipment which hasn't been built into the works at the time of invoice. At times these materials can be a significant cost which the contractor often has to carry for several months until they're built in. Many clients will however accept a bank guarantee or insurance bond to cover the costs of this material or equipment, thereby freeing up the contractor's cash flow.

Some clients only pay on completion of a milestone, and sometimes the contractor may have to complete a substantial amount of work at their cost in order to achieve the milestone before payment is received.

The client's budget

Some client's budgets are too low. Obviously contractors are not always privy to the client's budget but some research may give an indication of what it is.

When the budget is too low:

- The contractor's price is usually higher than the budget and the administration of the project is difficult because the client will argue and resist all claims and variations that the contractor submits even if they are legitimate and valid.
- The client will also try and save money wherever they can which could mean that they select cheap, poor quality materials which are difficult to use or which could cause a finish which is below expectation and frequently becomes the contractor's problem.
- Furthermore, the client may take short cuts with the design, or even reduce payments to the design team, which could affect the quality of the design information, or the effectiveness and timeliness of responses to queries submitted by the contractor to the team, which will impact the schedule and possibly result in delays to the contractor.
- Furthermore, when the project price is above the client's budget there is always the risk that the client runs out of money before the project is complete, leaving the contractor unpaid.

The quality of the tender documents

The quality of the tender documents is often indicative of the quality and experience of the client and their design team. Poor quality documents are an indication there could be problems in the course of the contract due to poor and insufficient information which will lead to delays and possible claims. In general I would stay well away from these contracts, or at least price them in the knowledge that you will have to use a well prepared and experienced contract team that can deal with any problems, and submit variations for the delays and inconveniences caused by the poor information. On occasion, if the client has a sufficiently large budget, and the contractor has an experienced team, it's also possible to profit out of the situation.

Some tender documents and scope of works are so poor it is unclear what the contractor should price so it's important that the contractor clearly defines all their assumptions and specifies what they have priced (or not priced) in their tender submission.

Liquidated damages

Some projects have excessively high or stringent liquidated damages. Often these damages may be uncapped, allowing the damages to quickly accumulate and consume any potential profits and even in some cases the entire value of the project. It's important to understand the liquidated damages and what the risks are of being penalised.

In some cases it's possible to qualify a lower value for the liquidated damages in the tender submission, as well as placing a cap on the maximum amount of damages that can be applied. If these proposals are not acceptable to the client it may become necessary to walk away from the tender. More than one contractor has gone out of business when they've not completed a project on time, resulting in financial charges and penalties which they were unable to pay.

The project schedule

As part of the liquidated damages it's also important to understand the project schedule and the project's completion dates. Often these dates aren't feasible in which case either propose a more realistic schedule or don't price the work. Rather, let another contractor be stuck with a problem project which they can't finish on time, resulting in liquidated damages as well as damage to their reputation.

Will you be competitive?

Often you know before submitting a tender that your price will be uncompetitive. The project may be too small, it could be located in an area where you aren't working, or you may not have the expertise or the equipment for the project. If the project is more suitable (either in location, size, or type of project) for other contractors your price is going to be far more expensive than theirs. Not only has valuable time been wasted in preparing and submitting the tender, time that could have been used more effectively on another bid, but there is also a risk the company's reputation has been damaged since the client may think the contractor is too expensive and uncompetitive and they will not invite them to price their next tender.

We've often declined tenders because we believed we would be uncompetitive. We usually phoned the client and sent them a letter, explaining why we weren't going to submit a tender and why we felt in this particular case we would not be competitive or have the necessary resources to offer them the quality of work they had come to expect from the company.

Illegal and criminal activities

Contractors should be careful not to be involved in illegal or criminal activities. Projects that are built illegally without the correct permits, or require vegetation to be cleared without the necessary permits being in place, can result in:

- bad publicity for the contractor since it's normally their machines and workers that appear on television and in media photographs
- the project being stopped with implications for the contractor's equipment and people
- equipment even being impounded
- the contractor not getting paid

Further to this, people often acquire money illegally which they use to finance building projects. While you may think there's nothing wrong with being paid from these ill-gotten gains there are always risks that:

- if the source of income is exposed the contractor receives negative publicity

- the client is arrested and the project comes to a halt
- the client's assets are seized resulting in the contractor being unpaid
- people involved in criminal activities may resort to illegal tactics should the contractor get into a dispute with them

Controversial projects

Sometimes certain projects are controversial, such as projects located in sensitive environmental areas, or projects that have resulted in local inhabitants being relocated. These projects may have vocal protesters who sometimes even resort to sabotage, vandalism and blocking off access to site. This opposition can delay the project, result in the company losing money as well as being in the press for the wrong reasons.

Summary

The right project to price is influenced by a number of important variables. Projects you should not be tendering on are ones which:

- are unprofitable
- pose a serious safety risky to workers or members of the public
- you don't have the expertise and knowledge to prepare an accurate tender
- have high risks that could affect the future of the company
- require staff with expertise and knowledge that the contractor doesn't have and will probably not require in the future
- the client may pay late
- will negatively impact the contractor's cash flow
- require bonds and sureties which cannot be obtained
- have a final value in excess of the client's budget
- have unachievable milestones
- have high and un-capped liquidated damages
- have unreasonable contract conditions
- have a weak design team
- don't have all the permits in place, or there's a chance the client won't be granted the necessary permits
- are too small
- are too large
- are in a remote region you have no experience with
- are for difficult clients
- are controversial and may tarnish the image of the company
- are illegal
- are funded by known criminal activities
- you will not have a competitive price giving you a fair chance of being awarded the project

Selecting and winning a suitable project for the company is vital to being a successful contractor.

Chapter 2–Finding the Right Projects

Many smaller companies are trapped in a cycle of doing small projects, working as subcontractors to other contractors, or working for difficult clients or ones that are poor payers. So how do you break out of this mould and find the right project with the right client? Well this is where you have to do some work to find who these clients are, and which projects are going to be good projects leading to further work.

Keeping a good client

Case study:

For many years we had been unable to work for a particular client and most of their work was awarded to two contractors. Eventually we got a lucky break and a fairly difficult project came up which wasn't that prestigious. One contractor was currently on site but hadn't performed particularly well, and to make matters worse they had recently transferred their Project Director to another project. The other contractor had also changed personnel within their organisation and their current management didn't understand the importance of keeping the client. We were desperate both for the work, and to establish a relationship with the client, so we bid keenly for the project and went all out to secure it in the post-tender negotiations. We were awarded the project and went on to successfully complete it as well as a number of smaller projects.

Following on from these we secured a very large project which we again successfully and profitably completed. By this stage the client had long forgotten any previous relationships they had had with either of the two original contractors.

The above shows how important it is to keep other contractors away from your clients and to ensure you look after them, since once lost it might be difficult to regain them again.

Of course you cannot keep competitors away from a client by price alone. There will always be someone who is prepared to undercut prices to get an opportunity to work for a good client. The only way to keep competitors out is by:

- developing a relationship with the client
- delivering a quality product, on time and without incident
- being fair and honest with the client
- acknowledging mistakes and rectifying the problems as soon as possible
- managing the client's expectations
- not over-promising and under-delivering – if anything under-promise and over-deliver

- senior management from the contractor being seen by the client on the project

If a competitor's price is cheaper than yours, explain to the client why your bid is more expensive. Remind the client of your past successes and achievements on their projects. Negotiate with the client, but don't undertake a project at a price that will lose money.

Naturally, when a contractor does repeated work for a client there's a danger they'll become complacent and arrogant, and start to charge inflated prices for work. When a client discovers they've been taken advantage of they can be quite unforgiving.

Don't become overly dependent on one client. Unfortunately clients also go through phases of spending and other times when they cut back on expenditure. Contractors who are largely dependent on work from one client can quickly run out of work when that client stops spending.

When work finally runs out with a particular client due to another contractor winning the work, or the client not having further construction projects, stay in contact with them. This may entail a phone call every six months and as a minimum a greeting card at the end of the year. No matter how good the relationship was with the client you cannot depend on them remembering your company in a year's time and inviting the company to tender for their next project. Regrettably clients have fairly short memories (except when it comes to remembering poor performance), so it's essential to keep reminding them of the past relationship.

Research

Large companies often announce new projects in press releases, so just reading newspapers provides some idea of projects starting soon. Local newspapers also have articles about new developments and planning permissions granted in the area.

Trade publications are useful in providing information about new projects. For example, mining industry publications have articles regarding new mining projects, and most industries have similar journals which are useful to subscribe to.

Reading newspapers and trade magazines makes you familiar with the names of clients who have finance and are starting new developments. Even advertisements provide names and contact details of managing contractors, designers and architects that work in the field that you're interested in, and can provide useful leads. Put together a data base of these clients, contractors and professionals.

Regularly check potential clients', designers' and managing contractors' websites since they often have announcements of new projects they've been awarded.

Always be vigilant in your travels and you'll notice old buildings being demolished, sites or estates being prepared for developments, and real estate agents' boards offering units for sale that are yet to be built.

Tender data base

In many countries there are companies that collect data on all the tenders that are publically advertised. They collate this information into lists which they regularly update and circulate to contractors who subscribe to their service. (It's important to note that many tenders aren't advertised publically and so probably won't appear on these lists).

Tender data bases are an essential form of learning what projects are available for tender. It's important to subscribe to at least one of those that operate in the region of your activity and ensure the tender data base you subscribe to covers all the projects and clients that are of interest to the company. Many of these data bases allow the tenders to be filtered according to region, type of contract or size of contract. Sometimes without the filters the number of tenders coming through every week can be quite overwhelming.

Bear in mind that even if the project is too large for the company, there may be an opportunity to contact the larger contractors pricing them who may want to subcontract a portion of the works to a smaller contractor.

Existing projects

While working on a project talk to the client's representatives who often have inside information regarding other projects the company is undertaking or may be starting. It's also useful to obtain contact details of people in their organisations who you can approach to get onto vendor or bidding lists. Add these names and contact details to your own data base since they could be useful in the future.

The designers, architects and managing contractors are also a wealth of useful information because they're often working on projects for a number of different clients. Some of these projects may still be in the early stages of development, and construction hasn't started.

Talking to subcontractors and other contractors can also yield new information and contacts.

Sometimes even suppliers have knowledge of upcoming projects because they've provided quotes to other contractors or supplied prices to clients to enable them to prepare their budgets.

Of course an existing project is the opportunity to demonstrate the capabilities of the company. It's important to deliver a quality product, on time and with minimal fuss and incidents. With luck this good work and professionalism will be noted and remembered by the client, their representatives and the project team, leading to further opportunities.

Ensure that existing projects have company brochures available on site to hand to prospective clients. Project Managers should have sufficient business cards with them at all times.

Networking

As mentioned above it's useful to network with clients, subcontractors, designers, architects and suppliers involved on the projects you're currently working on. Still, this networking should be taken further, so it's useful to belong to trade and industry associations or local business groups. It's often the case of not what you know, but who you know that leads to business success.

Sometimes even union representatives get to hear about new projects before they start because clients begin negotiating project labour agreements with them well in advance of a project starting.

Friends and relatives can also provide leads to work opportunities, as can people you meet at sporting events, functions and even on aeroplanes. Always carry spare business cards with you in case an opportunity arises.

Contacts

There are many other useful contacts who are often involved with a new project from its early stages such as:

- land surveyors
- quantity surveyors
- town planners
- environmentalists
- geotechnical engineers
- estate agents
- recruitment agencies
- local government officials

Employees

Don't underestimate the market knowledge of your own employees who often have friends in the industry working for subcontractors, clients, designers or other contractors, and who may learn of new projects long before they're reported. Encourage employees to report leads for potential work opportunities.

Of course employees should be encouraged to be salespeople for the company, promoting the company in a positive way. Employees must be aware of the capabilities of the company so that when asked by potential clients they are able to readily market the company.

Competitors

Be aware of where your competitors are working and for whom they are working. Competitors don't easily give away information but sometimes at informal gatherings or industry forums they may pass comments about their projects or clients.

Often competitors announce their new projects in press releases or on their websites.

Marketing

You must be able to sell your company and its capabilities to potential customers. Marketing is a continually evolving process but is largely dependent on the type of customer your business is aimed at and where they are located.

Sometimes one can approach specialist marketing companies, but these need to be treated with caution. It's easy to spend large amounts of money on marketing with few results to show for it.

Should you approach a marketing company, check that they have experience selling what your company does. Make sure that they clearly understand what your business is about, where you operate and who your customers are. Set a clear budget which is relative to the potential outcomes you're seeking.

However, I would advise that construction companies do their own marketing. This normally should take a multi-pronged approach since no one thing will work alone. It's often a trial and error approach, as what works for one company won't necessarily work for yours. In fact, what works for you today, may no longer work for you in a year's time, so it needs to be continually adapted and improved to suit the changing conditions. It's therefore important to understand what's working so you can discard what's not.

Website

A website is an essential ingredient to selling the company and many clients (even the small home owner planning renovations) will view a company's website. The website should be reasonably simple – in general avoid pop ups and videos (although these may be suitable for some specific companies). Clients want:

- a quick overview of the company's capabilities, what the company does and where it operates
- to know who to contact
- some examples (with photographs) of the projects the company has undertaken, with an indication of their size, what is was and who the client was
- to see registrations and certifications such as for quality, environmental and safety management
- (for larger contractors), to see copies of their financial results or the annual turnover for the last few years

There are various tricks that can be used to promote your website. But this is another topic on its own and is an evolving process.

Company brochures

Brochures are useful to hand to prospective clients since they give an overview of what the company's capabilities are and they can be kept and referred to at a later date.

- These brochures should be simple and take a similar format to the website, including contact details, a brief overview of the company and examples of recently completed projects.
- The brochures should look professional and must be printed on good quality paper.
- Take the time and trouble to proof-read the brochure carefully before printing to ensure the spelling, language and facts are correct. I often see poorly produced and written brochures which don't portray a good image of the company.

Photographs

Photographs as they say 'are better than a thousand words'. However if poor photographs are used in an advertisement, brochure or website they won't enhance the advert, and sometimes can even harm the company's image. Check the photographs to ensure:

- they are a good quality (photographs mustn't be grainy, out of focus or skewed)
- they don't show any unsafe acts
- the work shown is of a good quality
- the work site looks organised, neat and tidy
- they illustrate what you are trying to show – often photographs are used which show other contractor's work and it's difficult to see what was built by your company, or, the photograph is too small or taken from far away and the work isn't clear
- the photographs are relevant and show a diversity of work since I've seen artistic photographs in advertisements which don't display the contractor's abilities and resources
- plant and equipment in the photographs is clean, in good condition and preferably has the company's logo on it

Company logo

Logos are what people usually see first – these might be on items of equipment, sign boards, advertisements, business cards and letterheads. Now I'm not an expert on marketing so can only speak from my experience, but, I would recommend the logo is kept fairly simple with letters that can be clearly seen and distinguished. Think what the logo will look like on items of equipment, business cards and letterheads. Don't make the logo too long as it may become reduced in size to fit onto some areas. Stick to bold letters and colours, remembering that it will often be placed on items of equipment painted in different colours.

Advertising

Advertising can be useful but also costly with limited results. I worked for a large construction company and we were frequently being solicited by sales representatives from various journals to place advertisements in their publications.

- Many of these would have been a waste of money as their readers were unlikely to be our clients, or provide us with any work. So carefully consider who your clients will be and whether they will read the publication. Make sure the advertising medium reaches the right market and reflects the image you want the company to portray.
- When drafting an advertisement keep it simple but ensure it tells the story you want to tell. The reader at a glance should understand what your company does.

- The advertisement must have the name of the company and contact details – preferably website, email, telephone and physical address and the name of a contact person.
- The advert must attract attention. It's pointless paying good money for an advertisement that nobody will notice and read.
- The advertisement must not be too crammed and cluttered – some companies try and squeeze in too many pictures, or too much text, making it difficult to read.

Of course advertising takes many forms which will depend on the size of the company, the services offered and the target market. Advertising may include some of the following:

- placing adverts in the print media such as:
    - national newspapers
    - regional newspapers
    - local newspapers
    - magazines
    - journals
- placing advertisements in the electronic media such as:
    - online newspapers
    - online journals
    - websites
- other electronic formats such as:
    - Facebook
    - twitter accounts
- placing advertisements in strategic locations where they'll be noticed (always ensure that you have permission to put up the advertisements)
- delivering or handing out flyers which could be in the form of:
    - fridge magnets
    - writing pads
    - calendars
    - diaries
    - simple piece of paper

The important part of any of these advertisements is that they are directed at the market which you want to see and take note of your company. For example, if you're a local plumbing company that specialises in repairing leaks then it's pointless placing an advertisement in a national newspaper at great expense, rather, it may be more appropriate to hand deliver fridge magnets or note books with your company details to homes in your area so residents can find your contact details when a plumbing problem arises. By the same token electronic marketing is very successful for many products, and it's probably successful in certain fields of construction, but not in all.

It may be necessary to experiment to see what works and what doesn't. But even what works today may not work in a couple of years' time, so the strategy needs to be looked at periodically and modified to suit the current circumstances. Part of this process is always to ask new customers how they came to hear of you and these answers should be tabulated and reviewed to see what's successful and what hasn't worked.

Sponsorships

As part of their marketing campaign some companies sponsor sporting events, sporting teams or other events. From past experience I've seen little benefit for construction companies, but I'm sure some may disagree. On occasion, sponsorship of local teams and community events is useful in fostering goodwill in the community. Of course, with any sponsorship it's important the name of the company is given prominence and people understand who the company is and what services they provide.

Sometimes sponsorship is unavoidable such as when clients request a donation or sponsorship for an event. It's important once again that the name of the company is prominent at the event, which can take the form of setting up flags or banners.

Successful companies should consider contributing to charity, in particular to the local community where they've been successful. Preferably the donation should be something tangible and visible. Again, use and promote the opportunity by inviting the client, or even the local media, to handover ceremonies (always make sure both the project and donation are worthwhile).

Plant and equipment

Plant and equipment are often a good form of advertising the company so it should be kept clean and in good condition. The company's logo should be prominently displayed – consider including contact details. Of course drivers of company vehicles and equipment on public roads should obey the road rules and behave considerately to members of the public because you don't want the company to be noticed for the wrong reasons.

Even if the company has externally hired equipment it's possible to have the company's name on magnetic signs, or similar, which can be stuck to the equipment and then easily removed when it is returned to the hire company. (Of course, ensure that you don't damage the item's paintwork which will result in costs to repair the damage.)

Sign boards

Sign boards on your construction projects are an important form of advertising.

- Obviously seek permission from the property owner and from the relevant authorities before erecting the sign board.
- The signboard should be prepared by a professional sign writer and must be erected securely and properly. (Signs of a poor quality that are askew are a poor reflection on the company.)
- The signs must be clearly visible and legible and should have the company name and contact details.

Sometimes after the project is completed the contractor forgets to remove the sign board and they remain for several years afterwards which can be good free advertising as long as the project continues to be looked after and remains in good condition. Obviously, there's no guarantee that the client won't later use other contractors to carry out work which may not be as professional as your company's work.

Case study:

A company installed shade sails on a property with prominent street frontage and erected their sign board which remained in place for several years. However, with time the shade sails became torn, stained and mouldy. Yet, the sign was still there advertising the company – surely no longer a good advertisement.

Be cautious of the location of the sign boards as often these are on the property corner boundary, possibly being interpreted as relating to work being done on a neighbouring property which might not look as professional as you would like to portray the company.

Direct calls and meetings

An important part of marketing is visiting and meeting prospective clients. The meetings should be prearranged which means that you need to know who the best person is to talk to. I've regularly found that most people are happy to talk to you. However, if the meeting isn't with the correct person it can be a waste of time.

- Be prepared for the meeting. Study up on the company and their current and future projects.
- Take sufficient business cards and brochures to the meeting.
- Make notes of what's said.
- Introduce yourself, explaining your position and what your company does.
- Outline examples of projects the company has done which are similar to what they are involved with. Outline your company's experience, resources and strengths.
- Congratulate the person on any recent company achievements or new projects.
- Enquire how the existing projects are progressing.
- Enquire about new projects, the type of project, expected start date, estimated size, who to contact and how to be included on the tender lists.
- Explain why your company will be suited and experienced to work on the project.
- Offer assistance to the client with their feasibility studies and any questions they may have.
- Find out if there are other people in the organisation you should talk to.
- Provide additional copies of the company brochure and ask for these to be distributed to other people within the organisation.
- Thank them for meeting with you and send them a thankyou email which could include a link to your company's website as well as an electronic copy of the brochure.

Remember, large organisations often have a number of different divisions or operating entities, and sometimes these operate in their own individual compartments with little communication between them, so it may be necessary to make contact with each one individually.

Company newsletters

A useful form of marketing is to produce company newsletters.

- They can be a page long, or longer, depending on the size of the company.
- Quarterly letters are usually sufficient.
- Ensure they are proof read by a senior manager to ensure the facts are correct as well as the spelling and grammar.
- They should be distributed to staff and clients on existing projects.
- They should cover changes within the company and feature various projects.
- They should be produced on good quality paper.

Business cards

Business cards are an important advertising tool as they are handed out at business meetings, to potential clients, or at events.

- They should be uniform throughout the company with standard layout, logo and colours.
- If the company changes the layout, all business cards should be replaced. I find it a bit confusing when I go to a meeting and I get business cards that look different from people working for the same company.
- Since business cards are often pocketed and filed, when a person needs to refer to the card they often can't remember what the company was. Each card should therefore clearly have the person's name, title and contact details, and the company name, which should reference what the company does, for example; building contractors, electrical contractors and so on.
- The cards shouldn't be cluttered and should be able to be read by most people without using glasses so it's important to carefully consider background colours, designs and logos.

Referrals

Referrals are really an excellent form of advertising – after all someone else is doing your marketing for you. Wherever you work, try and hand out business cards and brochures. For instance, you may be a tiling company working for a builder renovating someone's house. Hand out your card to the owner and any neighbours you see; you never know, they may have other tiling work which they want you to do – even if it's only much later. In fact give them a couple of spare cards. I know we've often recommended good tradespeople to our friends and passed on the contractor's business card.

Entertainment

Some clients expect to be entertained or invited to sporting events, though many clients have strict rules governing their staff attending such events. Ensure you understand and follow these rules.

Entertaining clients can often be a waste of money and can even do more harm than good. Be selective with who is invited, both from the client's side as well as from within the company. There are different forms of entertainment, from individual project events with the personnel immediately involved on the project, to larger corporate events where senior members of companies are invited. Entertainment must be appropriate.

All formal project and corporate functions should be carefully thought out and planned.

- Senior managers need to review the guest list to ensure no one who should have been invited has been left off.
- Formal invitations should be sent out.
- The event must be properly organised since events with insufficient food, drink or seating can be embarrassing.
- These events shouldn't be organised too frequently because people stop attending when the novelty wears off.

Sources of work

Personally I am reluctant to work as a subcontractor to a larger contractor and would much rather work directly for the client, but this isn't always possible. In addition, it can be useful for smaller contractors to work as a subcontractor which could provide them an opportunity to meet the client and to demonstrate their capabilities. Major contractors will try and keep their subcontractors away from direct involvement with their client, but being on the project usually provides opportunities to meet the right people. Make sure all your personnel have clearly marked clothing and that equipment has your logo prominently displayed.

Additionally, these projects give smaller contractors experience of working on large projects for major clients. This experience eventually assists with building a good resume of projects, and can be used to convince future clients of the company's abilities.

Working on these larger projects is also an opportunity to learn new skills and to observe and learn different construction techniques.

It shouldn't be overlooked that even projects which aren't necessarily related to the field that you work in, and which you would not tender for, may have a small portion of work in your field. For instance large electrical contractors and piping contractors might not be interested in performing the excavation works and minor concrete works and would subcontract these out to small civil contractors.

Registrations with trade bodies

Even when it's not compulsory it can be useful to register with the relevant trade bodies, or organisations representing the industry the contractor operates in. Many clients perceive this as providing them with a guarantee that the companies registered are more reputable than others who aren't.

Furthermore, clients often contact these organisations to ask for names of suitable contractors in their region.

Sometimes these organisations have newsletters which can provide useful information or contacts. They also usually have regular meetings with their members which provide opportunities to meet competitors. Some even provide training programs which members can attend.

Centralised marketing and tendering (bidding) organisations

In some areas there are organisations set up that clients can contact to find local tradespeople. The client phones or emails the organisation and gives details of the job that needs doing. The organisation then approaches plumbers registered on their data base, who have the appropriate experience and that service the location of the project, to quote for the work. The plumber either quotes directly to the client or provides the organisation with their price for them to pass onto the client.

This can be useful for small contractors that provide a service for the average home owner, since the organisation effectively markets the company. However, there's a fee to be paid, which is either monthly or dependent on the leads the organisation provides, or the value of projects the contractor wins.

It's also difficult to differentiate your company from others except by being the cheapest, and by slowly building a reputation of delivering quality work on time.

Forming joint ventures

One way of taking on bigger projects is by forming a joint venture with another contractor. Joint ventures have a number of benefits:

- the combined resources of the two partners enables them to tackle larger projects
- they can be formed when one partner has a particular expertise that the other partner doesn't
- one partner may have a relationship with a client but not the capacity to undertake the project on their own
- if both partners were going to tender for the project it effectively reduces the competition to win the project

Not only do the joint ventures provide the companies with exposure to working on larger projects, but they also provide opportunities of seeing how other companies do business and to learn new techniques and systems.

Partnering with local contractors and communities

Sometimes when the company is planning to work in a new area it's useful to form a partnership with a local company. In regional areas, particularly with large indigenous populations, it may be worth developing a relationship with the local community.

This strategy can be particularly worthwhile ahead of the start of large projects. Clients are often under pressure to support the surrounding communities and companies, but these local companies frequently have insufficient experience or resources. Establishing relationships (either in the form of a joint venture or committing to subcontract portions of the work to local contractors) makes the tender more attractive to the client. In addition, these contractors assist with their knowledge of the area. Using local resources is also usually cheaper than bringing them in from elsewhere.

Of course, these relationships should also benefit the local companies and communities. At no time must it seem that you have cheated the local companies because the client will almost certainly hear of any problems.

Political support – understand who can actually help

Some companies develop relationships with local or national politicians or groupings. These relations can be useful in some regions or countries, particularly when the relationship is used to obtain information about forthcoming projects and developments. Nevertheless, I would be cautious of some associations since political figures come and go, and their replacements might not want anything to do with the friends and perceived allies of their predecessors.

In some regions these political friendships can quickly develop into corrupt relationships and dealings which should be avoided.

Corruption

In some countries bribery and corruption is rife and it's best to avoid these countries. There's no guarantee that the official receiving the bribe can actually influence the awarding of the project, or that they aren't dealing with other contractors at the same time. In addition, even in countries where corruption is prevalent bribery is normally an offence, and a company caught paying a bribe may find their management arrested and the company barred from doing further work in that and other countries.

If you believe that corruption could play a role in the tender process, or the awarding of a project, then rather decline the tender.

Unfortunately corruption isn't just about paying off officials who can influence the bid, but sometimes one contractor may pay another contractor not to tender or to submit a higher non-competitive tender. These practices are illegal and if contractors are caught it will severely harm their reputation, even leading to their management being imprisoned or fined.

Design and construct projects

Some projects are put out to tender as 'design and construct' projects so it's useful to have a relationship with designers to enable the company to tender for these. Be aware that tendering for design and construct projects can be more costly and time consuming than other tenders.

There are also more risks associated with these projects and it's important to work with designers who have experience of designing similar projects. Also, ensure that they have sufficient design indemnity insurance in case there's an error with their design.

Multi-disciplinary projects

Some clients put projects out as complete packages, including civil, mechanical and even the electrical works. Companies that have all of these components in-house usually have a competitive advantage. Alternatively, the company could form partnerships with suitable contractors who can provide the needed expertise which the company doesn't have, forming either joint ventures or alliances for the project.

Finance

Sometimes clients have projects that are viable but they don't have the finances in place to proceed. Contractors in the fortunate position of having a strong bank balance, or a source to borrow capital, may consider offering the client a finance option to make their tender more appealing. In addition, a well-structured and favourable finance deal could net the construction company additional profit.

Obviously care must be taken to ensure there are sufficient guarantees in place to cover the loan amount and that the value of the work doesn't exceed the agreed loan value.

Assisting with feasibility studies

Often clients approach contractors to assist with their feasibility studies. There may even be opportunities to approach a client when you first hear about a potential project to see if you can assist them in any way. These feasibility studies do take time and effort and some projects may never eventuate, or when they do it's several years later when you are no longer interested in the project.

However, when a client requests a contractor to assist with their feasibility study it's sometimes difficult to refuse. I've had clients not invite contractors to tender because they had declined to assist them with their feasibility study.

By helping with the feasibility study it's possible to influence the client's design, steering them in a direction better suited to your company's abilities. Also, being involved with the project's feasibility study gives the contractor a head-start in the tender process since they've already put thought into the construction methodology and researched various options which should lead to a more accurate and competitive tender price.

Beware of overly ambitious projects

Over the years our company was often approached by developers to assist with feasibility studies for projects they were proposing that were never going to be built. The projects were simply too ambitious or outrageous for the location, they weren't going to get finance or they weren't going to get development approval. Very occasionally, maybe one out of ten times, the project went ahead, but usually several years later when the developer had long forgotten who assisted them with their planning and feasibility studies.

Assisting with these projects is time consuming and takes resources away from other projects which are more likely to proceed and provide work in the short and medium term. If we had priced all of these schemes we would have had to increase the number of Estimators in our company.

It's therefore important to research and understand which projects are real and which probably won't happen, or if they do, when they will happen.

If there really isn't anything else for the estimating department to do, you can consider assisting with these projects.

Sometimes these projects are brought to you by good clients who provide the company with other work, so it's then necessary in the interests of continuing the relationship to assist.

Tracking potential projects

It's essential to track the various potential projects since many could be several months or years away, and often their start dates change. Prepare a list which includes the name of the project, the details for the contact person, the anticipated start date and the value of the project. This list should be divided into projects which will be out to tender shortly, medium term prospects and, lastly, those which are some-time away.

Continually update the list and maintain contact with the clients to ensure that when the project is put out to tender you will be invited to tender.

By tracking potential projects it's possible to decide which the most suitable ones are to pursue. The number of potential projects may also influence the profit margin used when tendering, since if there are only a few prospective projects it may become imperative to secure the project. However, if there are many potential projects there will be other opportunities if your bid is unsuccessful so you can possibly add a larger profit to your price.

Mailing lists and data bases

Contractors should maintain a data base of potential clients, engineering companies, managing contractors and architects. These lists would include the company name and contact details of important persons within the organisation. These data bases are important when sending out invitations to company functions, or greeting cards at the end of the year. All senior management, and in particular project managers, should contribute and update this list.

When reading newspapers and trade publications you may come across mention that someone on the list has been promoted or moved to another company. Use this information to not only update the list, but also to contact them, offering congratulations and good wishes in their new position.

Often companies win awards or make the news for other reasons. Again, use these opportunities to maintain contact by sending out messages of congratulations, or even condolences should the company have suffered an accident on a project.

Ensure these messages are sent by a suitable member of the company and that the messages aren't just generic letters – a handwritten note can often generate lots of good will.

Business Development Managers

Many larger companies have Business Development Managers (BDMs) who are tasked with finding new potential projects and clients. These can be useful since they can devote time to finding new projects and contacts, developing relationships, and preparing marketing material. Most companies however can't afford to add an additional person to their overheads.

Many BDMs need to be steered in the direction in which the company wants to head. I've seen these individuals chase down clients and opportunities that weren't suitable for the company. After pursuing a client for a tender and finally getting the opportunity to bid, it's awkward if the project isn't suitable and the contractor has to decline pricing it.

Some BDMs aren't familiar with either the construction industry or the company's abilities so it's essential for senior managers to review the contacts and relationships they're developing, as well as the marketing material they're preparing and distributing.

Trade registrations

With certain trades, or in certain industries or regions, it's imperative to have the required qualifications and registrations and failure to have them could mean that the contractor is unable to tender for some projects.

Quality, environmental and safety accreditations

Larger clients expect their contractors to have the appropriate certifications in place proving they comply with the standardised quality, safety and environmental systems and procedures. Even if this isn't specified it's almost certainly something that clients will check, and it may prejudice the contractor if they are not in place.

Green building and sustainability

In many countries there's a growing market for constructing buildings which are environmentally friendly. A large part of this is in the design of the building which in most cases the contractor doesn't control. There is however merit in teaming up with architects and designers who are able to design environmentally friendly buildings so that you can offer a total package.

Part of what makes a building environmentally friendly is the construction process and the following should be considered:

- Recycling and reusing building rubble.
- Segregating waste.
- Using environmentally friendly materials which depends on:
    - their manufacturing process
    - the distance they have to be transported to the project

  - how easily they can be recycled at the end of the life of the facility
  - the amount of wastage
  - toxins they may emit during their life and particular during construction
- Ensuring that suppliers and subcontractors comply with environmental legislation and procedures.
- Having environmental plans, policies and procedures in place.
- Having workers and staff committed to achieving the highest environmental standards.
- Eliminating wastage of power and water and minimising their use during construction.

In some regions there may even be accreditations that can be obtained that certify the company is competent to build green buildings.

There are other opportunities to consider in the green field, particularly when it comes to renewable energy where there may be construction opportunities. For example, in solar and wind energy the contractor could either team up with one of these providers or look at becoming a niche contractor in these fields.

Maintenance, refurbishments, repairs and renovations

Many first world countries are no longer building new infrastructure or facilities, but, large portions of the existing structures are thirty or more years old and require replacing or major maintenance. Therefore there's good potential for work in the rehabilitation, repair and renovation market, and good contractors may have unlimited work.

To become part of this market, consideration should be given to learning about new repair techniques and products. Much of the existing infrastructure is below ground, so being able to repair or replace this without digging up city streets could provide a steady source of work.

Innovative contracting or tendering methods

It's possible to make the company more competitive or more appealing to the client by proposing alternative contracting techniques such as:

- design and build
- partnerships with the client
- tying up specialist suppliers or subcontractors in a tender consortium or joint venture
- entering into a target plus contract with profit sharing where both the contractor and the client benefit from innovations in the methods of construction

A good reputation

Reputation is discussed in more detail in chapter 13. Having a good reputation often means that clients actively seek the contractor out, automatically inviting them to tender. Nonetheless, if the contractor has a poor reputation no matter how much effort is put into trying to work on a project, or with a client, it will be doomed to failure.

Testimonials and references from happy clients are very useful for marketing the company.

Saying no

Sometimes clients request contractors to undertake projects which are either unsuitable for the company's expertise and experience, are too large, too small, that are difficult or are required at a time when the company simply doesn't have adequate resources available. In these cases you have to learn to say no. This should be done both firmly and politely. I've gotten into trouble undertaking projects that I really didn't want to do for the reasons above. Some of these projects resulted in us losing money, but worse, a couple have actually harmed our reputation because we had to use inexperienced staff on them. The harm done to client relations was usually worse than if we'd simply declined the project in the first place. This doesn't mean you shouldn't help your good clients where possible, even if it means undertaking a project at a lower profit. However, never be forced into a project that will harm the company's reputation or cause it to lose money.

Be honest with the client, giving your reasons for being unable to price or do the work. If possible make suggestions as to when or how the company could be available to do the project.

Summary

It's important for all contractors to find suitable projects on which to tender. This involves:

- research such as reading newspapers, trade and financial publications and checking clients', designers' and contractors' websites
- subscribing to a tender data base
- marketing such as:
  - establishing a website
  - producing a company brochure
  - printing and distributing business cards
  - distributing company newsletters
  - appropriate advertising
  - contacting and meeting potential clients
  - erecting project sign boards
  - ensuring logos are on company vehicles and equipment
- getting clients to refer your company to other clients, project management teams, industry professionals, their friends or neighbours
- using opportunities on existing projects to talk to clients, suppliers, designers and subcontractors about new business opportunities
- belonging to centralised marketing organisations
- networking
- asking employees for leads they may have for potential projects
- talking to other professionals such as surveyors, geotechnical engineers and town planners

- asking friends and relatives about their possible contacts
- establishing a data base of potential projects and clients
- looking after and maintaining existing relationships with good clients
- forming joint ventures with other contractors that may be local, could increase the size of project that can be tendered for, or bring in another level of expertise
- coming up with alternative tender strategies
- developing alternative techniques or knowledge that will enable the company to enter a niche or developing market
- ensuring the company has a good reputation
- assisting clients with their feasibility studies
- engaging political support which sometimes can be useful, but always taking care to ensure this support is genuine and at all times avoiding corruption, bribery or other impropriety
- having the appropriate registrations and accreditations in place
- being able to tender for design and construct projects and projects which are multidisciplinary in nature
- even offering to finance projects if this is feasible and appropriate

Sometimes it's necessary to decline a project which isn't suitable or convenient at that time. Care must be taken not to upset the client in this process.

Chapter 3 – Tendering (Preparing the Quotation)

I've worked for some companies that submitted dozens of tenders every month without giving too much thought to any of them. We weren't selective and tendered for everything. Many of these tenders were unsuccessful even though we actually needed the work. The estimators worked long-hours and were frustrated with the high failure rate. Furthermore, the tenders were done in such a hurry that errors often occurred. Often our price was too expensive meaning we didn't win the project, while other times our price was too cheap and we were awarded the project at a price that was too low and lost money.

Frequently the hurried nature of preparing the bid was apparent in our tender submissions. These were a poor reflection on the company and its capabilities and our tender was often discarded by the client even when we had the lowest price.

Tendering is one of the most important functions of running a construction business. If it's of a poor standard, with a price that's too high, the company won't be awarded work and will go out of business. On the other hand mistakes could result in the contractor being awarded a project at a price below what it will cost to build, the schedule may be unachievable, or the contract conditions have been misunderstood, all of which could lead to the company losing money.

It's important that the person responsible for preparing the tender:

- is diligent, checking the tender submission carefully for errors and ensuring it complies with the client's tender documentation
- understands the tender system the company is using (this is often a propriety tender package system)
- understands the methods of construction (it would be pointless to, say, have a person familiar with electrical projects doing a tender for a road)
- is aware of the company's available resources and its capabilities
- understands the rates of productions, both for personnel and for equipment (these rates will differ between countries, regions, industries and even projects – for instance production rates will vary between first world and third world countries, they could be lower working within an existing facility or in a controlled petrochemical or mining project compared to say on a building project in the city)

Case study:

One company I worked for employed someone to do tenders who had never worked on a construction project before and had been elevated from the position of general office person. Needless to say this was a recipe for disaster because he did not have the experience or knowledge to understand how things were built. We had to redo the tenders after he had finished them.

Mistakes made during the tender process can be extremely costly.

Case study:

One of my projects was the civil and earthworks for a new gas power station valued at twenty three million dollars. We were a subcontractor to the managing contractor.

The tender documents were poor and the managing contractor had merely issued us the documents they had received from the client that included the scope and specifications for the whole facility including the mechanical and electrical works. There was no separate scope for the works we had to do. The only drawing we were given was an overall footprint of the facility showing the outline of various structures, and a contour plan indicating the surrounding contours outside the perimeter of the plant.

We were verbally told to allow for eighteen hundred cubic metres of concrete, but had to guess the quantities of formwork, reinforcing and underground pipework.

Using the contour drawing we estimated the quantity of earthworks required to construct the power plant terrace. Because the contour lines stopped at the edge of the plant the Estimator had to assume what the topography would be under the footprint of the plant. During the tender process we were issued another layout drawing which showed the footprint reduced by 40%, so we based our earthworks quantities on this drawing. There was no record that we were issued this drawing.

Now clearly there was insufficient information to price such a complex job, and the little information we had was of a poor quality, some given verbally or via 'back-door' drawing issues. Yet the company appointed an inexperienced Project Manager to price the work. Somehow he put together some semblance of a bill of quantities and priced them. He then submitted a lump sum price to the managing contractor. The tender submission had no reference to the assumptions made in the tender, or to what drawings were used to arrive at the price.

Starting the project I immediately found there were a number of problems with the tender which included:

- *our tender only allowed for the services and facilities required for our work, but the managing contractor felt we should have priced all the requirements in the document for the whole project, so for instance, they felt we should have allowed to provide site offices and facilities for them, the client and other contractors on site*
- *after we surveyed the site we found that our assumptions of the contours were incorrect and there was a substantial increase in the earthworks quantities*
- *our assumptions on the types and quantities of underground pipework was incorrect*

These together with other errors added nearly two million dollars to our costs before we even started the project.

The lesson from the above is to ensure that the tender submission:

- qualifies what's been included and what's excluded
- explains the assumptions made in order to arrive at the price
- should note what drawings and information were used to formulate the price (especially if revised drawings are issued during the tender process)

Case study:

One company had the Project Managers pricing various small projects. This was uncontrolled and many of these projects lost money. On investigation I found that the projects weren't being priced correctly. I instituted a review process to ensure that we only priced projects which we wanted, that they were priced in a consistent manner, and that the pricing was accurate. With better control of our tenders we were more successful with our projects.

Tendering should not be taken lightly. I recommend that companies set a limit on the size of projects that the estimating department can submit without senior management reviewing the tender.

Tenders cost money to prepare so it's important that this money is well spent and companies concentrate only on tenders they want to win, giving every effort to do this at the best price.

Check the tender documents are complete

When tender documents are received it's important to check them to ensure they are complete and include all the pages, sections and drawings. I've often received documents missing pages or sections. On occasion drawings on the drawing list haven't been included or the revision numbers were different.

When a tender is submitted the client assumes it's based on their documentation and all the information supplied.

Failure to check the documentation can lead to expensive errors if an important specification or drawing wasn't received and taken into account in the estimate.

Read the document

The Estimator should carefully read through the tender documents and make notes as they proceed.

It's important to note the tender requirements such as:

- When the tender must be submitted. A late submission is usually disqualified.
- Where the tender should be submitted. If this is some distance away the tender may have to be ready the day before so it can be couriered to the tender receipt location. I've heard of tenders being delivered to the wrong address, consequently not being considered by the client.
- The numbers of copies required.
- The format of the submission.
- Specific documentation or attachments required such as:
    - safety information

  - quality documentation
  - schedules
  - method statements
  - company financial statements
  - guarantees
  - insurance policies

  (Some of these documents may take several days or even weeks to put together and may involve input from other departments. Allocate the preparation of the various documents to a responsible person and track their progress to ensure they're ready before the due date.)
- Summarise important items like:
  - the name of the client
  - the client's agents and designers
  - payment terms and conditions
  - insurance requirements
  - general contract conditions
  - special conditions
  - liquidated damages
  - milestone dates
  - the basis of the tender such as lump sum or re-measurable
  - the estimated value of the project or a brief summary of the important quantities
  - a brief description of the project
- What the client will supply.
- What the contractor should allow for.
- Specific project rules and conditions, including project labour agreements.
- Other requirements and conditions which may affect the cost of the works.
- Any discrepancies in the document or points that require clarification.

It's useful to prepare a standard form that can be used for this summary.

Basis of tender

Projects are put out to tender in various forms.

- A cost recovery basis is where the contractor submits a schedule of resource rates and is reimbursed for the hours that their personnel and equipment work on the project. These projects carry very little risk for the contractor providing the rates submitted take into account all the costs the contractor will incur in carrying out the work, and providing the contractor records all the hours worked on the project.
- A re-measurable contract where the client supplies an estimate of the quantities in the tender and the contractor prices this schedule or bill of quantities. As the project proceeds the quantities provided may vary and the contractor may end up doing more, or less work, than was originally tendered for, and the final contract sum will vary accordingly. The changes in quantities may affect the schedule and the contractor's

overheads. It's important that the Estimator takes into account possible changes and ensures the contractor will be adequately compensated.

- A lump sum contract, meaning the contractor must during the tender phase calculate all the quantities required to construct the project and then price them accordingly. The contractor takes on the risk for incorrectly calculating these quantities. In some cases the contractor may elect to add in a contingency to cover any errors in their calculations.
- A design and construct contract, where the contractor is responsible for designing and constructing the project. Obviously at tender stage most contractors won't want to spend money on a detailed design and, in fact, there is usually insufficient time to do so. Generally only a preliminary design is done to enable a basic price to be determined. The contractor then adds a contingency to their price to allow for the incompleteness of the design. The more detailed the design the smaller the contingency can be.

Understand the project

The Estimator must understand what is included in the scope of the project. Most tenders include a scope of works and this should be compared to the drawings as well as to the client's schedule and client supplied bill of quantities. It may be necessary to query details of the project with the client or to request further information if the scope is unclear or if there are inconsistencies between the scope, drawings and schedule. If the client is unwilling or unable to answer some of these queries then the Estimator may have to make some assumptions to enable the tender to be priced. It's important that any assumptions made are included in the list of qualifications submitted with the tender.

Understand the contract

The Estimator should note the particular conditions of the contract document and how the risks are apportioned between the client and the contractor. Check whether the conditions are standard or if the client has added their own conditions, some of which may be unacceptable to the contractor. Some of the clauses may have to be discussed with senior management to see if they're acceptable and whether they'll be exposing the company to unnecessary or unwanted risks. In some cases it may even be advisable to seek specialist legal advice.

Laws governing the contract

It's essential to understand the laws governing the tender documents. Often these laws are of another country with which the contractor might not be familiar. The contract may also specify that disputes are dealt with in a foreign country which would add additional costs and risks to the contractor should a dispute arise. If possible the tender should be qualified so that any disputes are handled and resolved in the contractor's home country, or where this isn't possible, in one that has a similar legal system.

Site visits (site inspections)

Whenever possible the project site should be viewed during the tender process. This often takes the form of a client organised formal compulsory inspection, or it may be an informal optional visit. Either way the visit is important and can provide much useful information. If possible the person responsible for the tender should visit the project site, but this isn't always possible since they may be too busy or the project could be in a remote location.

Case study:

We sent a young engineer to attend a site inspection. When we came to review the tender we called the engineer in to explain to us what the site looked like. It was most frustrating as the engineer couldn't give us any useful information about the site. I was quite annoyed, but in hindsight we should have given her a better briefing and explained what to look for and note during the site visit.

When someone, other than the Estimator, attends a site inspection, it's important that they remember they are acting as the eyes and ears of the person doing the tender and they need to note as much useful information about the project site and its immediate area as possible, as well as who the other attendees and the client's representatives were. The quality of the information provided could substantially influence the way the tender is completed and may end up helping to win the tender or contribute to losing it. Potential problems which are missed during the site inspection could later be costly to the company because they weren't taken into account when formulating the tender price or in preparing the tender schedule.

As part of the visit it's important to be prepared.

- Know where the visit will take place and obtain directions.
- Know who to meet and ensure you have a contact telephone number so you can telephone should you get lost or be running late. (However remember that if the contact person is conducting the visit they'll probably switch their telephone off once the meeting begins and won't be contactable.)
- First impressions are important and if the contractor's representative appears disorganised, arrives late or becomes lost, the client will assume this is typical of the contractor's employees.
- It's useful to arrive early because this can provide an opportunity to talk to the client or their team. An informal discussion can often reveal some valuable additional items of information that other contractors might not be privy to. Arriving early also provides an opportunity to chat to other contractors and they could reveal if they are working in the area, and any concerns they have regarding the project as well as their level of interest.
- Have the correct personal protective equipment since it'll be embarrassing arriving without it and having to borrow from the client. If no equipment is available it might be impossible to visit certain areas of the project, or in the worst case it could even mean the project site

cannot be visited at all. The safety equipment should be in good working order and you should appear neat and tidy. Where the requirements are unclear contact the client to find out what they are. In fact, it's sometimes a good idea to telephone the client's representative ahead of the visit to ask some questions because this provides an excuse to introduce yourself, and provides the opportunity to get the contractor's name registered in their mind.

- In many cases the visitor may need to present a form of identification to access the site. In certain circumstances the client may require your details ahead of time.
- Take a notebook and pen. Even if the client says they'll distribute formal minutes of the site visit meeting, I would recommend that notes are taken of what's discussed at the meeting, since often the formal minutes are brief and don't cover everything, and they're usually only distributed several days later.
- If it's a formal compulsory visit there's often a form in the tender document that the client must sign, so this should be brought to the visit.
- Part of the preparation should include looking at the document and drawings so you have an understanding of what the project entails. Checking these beforehand may perhaps give rise to questions to be raised during the inspection. It's a good idea to take a copy of the overall layout drawing to the visit to help with orientation.
- If possible take photographs of the project site, providing this is allowed.

During the visit note the following:

- The distance the project is from Head Office (or the company's nearest office).
- The nearest towns and infrastructure.
- The name of the other contracting companies attending the meeting because it's useful to know who is competing for the tender. Note who their representatives are since the number and seniority sometimes provides an indication of the importance of the tender to them.
- The names of the client's representatives and their roles.
- The names of specialist subcontractors attending the visit because the Estimators may need to contact them to obtain prices for portions of the project.
- The conditions on the site such as:
    - is it open or restricted
    - are there other contractors on site
    - is the site flat or steeply sloping
    - is the ground soft or hard, and if there's rock what is its depth
    - is there ground water present and at what depth
    - are there existing structures that will impact construction
    - the site security and the level of security which may be required
    - is the site fenced
    - are there restrictions to get vehicles and people onto the site

  - the location of the laydown areas
  - the availability of services and utilities, their location, other contractors using them and any restrictions on their use
  - access to the site such as road condition, gradients, load restrictions and limitations
  - traffic conditions in the area
  - the location of rubbish and spoil areas
  - if your project involves earthworks, the location of the fill material, how easy it will be to extract, and its quality
  - the condition of the site roads and any restrictions
- The other contractors working on the site or in the vicinity, and the type of plant and equipment they have.
- Questions asked during the meeting and their answers.
- Note what the client's representative emphasises since it's usually important to consider these points, and highlight them in the tender submission.

Sign the attendance register and include the contact details of the lead Estimator. (In the past I have had young engineers attend site visits and leave their email address. They were then sent all the tender correspondence which they sometimes didn't forward onto the estimating team.)

When the formal part of the visit is over stay behind to hear what the other contractors are saying or asking, or what ideas they could have. On occasion another contractor already has a good relationship with the client and it's useful to note this.

After visiting the site explore the surrounding area and check for:

- Possible suppliers and contractors, noting their size, equipment, facilities and contact details.
- What accommodation is available and the distance from the project, if this is relevant.
- The transport networks to the area.

Prepare a brief report on the visit, including all relevant information and photographs, and submit this to the estimating team.

It's useful if the company has a standard document or checklist that can be used and filled in during site visits.

Supplementary tender documentation

The client's tender documentation may make reference to supplementary documentation which possibly isn't included in the documents. This could include reference to the client's insurance policies, standard specifications and geological reports amongst others. It's often important for the Estimator to read through these documents, so it's necessary to request copies of them or read them at the client's offices if this is all that's possible.

Tender schedule (program or programme)

This is one of the most important processes in preparing a tender and yet it's something that is often overlooked or done poorly.

Sometimes the client provides a schedule in the tender documents. Nevertheless, this schedule may well be incorrect or unachievable and many are just a guide. It could even have been produced by someone in the client's team who isn't even familiar with the construction processes. In many cases these tender schedules are far too optimistic. Consequently it's essential that the contractor ensures they can meet the client's schedule and if necessary propose an alternative one.

It's important to note that by accepting the client's milestone dates or schedule in the tender the contractor is committing to them. Failure to meet these dates during construction will probably lead to penalties and additional costs to the contractor. The contract schedule will have to conform reasonably to the tender schedule unless the scope has changed.

In many cases contractors prepare and submit their own tender schedule but often little thought has gone into it, it's been done in a hurry and may have been prepared by an inexperienced Estimator or Planner. The schedule may not have taken into account the site conditions, rules, or restrictions on working hours, types of machines or resources that can be used.

An accurate schedule is usually an essential aid to price the tender so it's vital that it's prepared properly.

Quantities

In order to price the tender correctly it's necessary to know the quantities involved for each individual task. The contract may be re-measurable with the client supplying a bill of quantities, even so, it's good practice to, where possible, check the larger quantities which could influence the schedule, or the final contract sum, should they be hugely different from those provided.

The client might request a lump sum price in which case it's the contractor's responsibility to measure the works and draw up a bill of quantities. Obviously it's important to ensure these quantities are accurate in order to obtain the correct price. The quantities should be broken down into sufficient detail so that the items can be priced.

Tender calculations

There are many different ways of working out the estimated cost of a project. Many smaller contractors look at a project and guess that it's similar to one of their previous ones and use that price with minor modifications taking into account any obvious differences between the projects. Clearly this isn't very accurate and doesn't really take into account variations in the quantities or the complexity of the project.

Ideally the Estimator should prepare a list or schedule of all the tasks or items required to construct the project. Each task should then be individually priced. These amounts are then added together to get the total cost of constructing the complete project. The overhead costs are then added, together with the mark-up and contingency allowance (see below).

Many contractors are inclined to use standard rates from previous contracts when pricing the items, which isn't very satisfactory since these vary from one project to another, as do their production rates. Material prices also vary from project to project. For example, the rate to place one cubic meter of concrete on one project will almost definitely not be the same on the next one. By breaking the items into labour, material and equipment it's possible to take these variations into account when pricing them.

The pricing can be in the form of a simple spread sheet or it could take the form of a computerised tendering package. There are many different tender packages with some being better than others. Many contractors select the cheapest package which may not necessarily be the best option for the company.

Each tender should have the hourly cost worked out for the individual labour trades. These rates must include the base labour cost, plus a factor to account for the estimated overtime hours, as well as allowances for leave pay, sick leave, bonuses, unproductive time, retirement funding, public holiday provision and other monetary allowances. In addition, somewhere in the tender add in the costs for protective clothing, mobilisation of staff and workers, training, as well as transport and accommodation costs. (Some companies choose to add these into their task labour costs.) On some projects the unproductive allowance can be quite large to allow for workers moving between tasks, the preparation of work method statements, attendance of safety and tool box meetings and the interruption of work due to the client's or other contractor's operations.

By working out the size and composition of the team required to complete a task, the production that will be achieved per hour, and knowing the individual hourly rate (calculated above), it's possible to work out the labour cost to complete each task.

When working out material costs it's important to allow for any wastage, bulking and compaction, laps, transport and handling, as well as the cost of fixings.

It's essential to diligently and accurately work out the costs since simple arithmetic errors can have serious consequences.

Ensure that the price for each task includes all of the activities required to complete the task. For example, when pricing an excavation it's not only necessary to include the cost of excavating the footprint of the structure but often it's necessary to include excavation for working space, battering of the excavation to make the sides safe, access and ramps into the excavation, protection of the excavation, and backfilling the excavation, working space, batters and ramps. Often the volume of earth removed from the footprint of the structure is only a small part of the total volume that must be excavated to construct the structure.

Calculation of overheads

There are two types of overhead costs on a project.

- Firstly those related to the operation of the company and the services the company provides to the project. These include the cost of running the company Head Office and the costs of the company's management and administration staff. Many companies charge each project a fee to cover these overheads and this is sometimes incorporated into the tender price. The Estimator needs to understand the company policy in this regard so that the correct amount is allowed for.

- Secondly there are the overhead costs directly related to running and administrating the project. These costs cannot be attributable to a particular item or task on the project and may include costs such as:
    - The project offices and facilities including their transport and set-up.
    - Management, support staff (Administrators, Planners, Engineers, Safety and Quality Advisors) and supervisory personnel. These costs would include their basic pay as well as provisions for bonuses, leave, sick leave, retirement benefits, allowances, other benefits and possibly also the cost for their accommodation and transport.
    - Mobilisation costs for personnel and equipment.
    - Specialised equipment such as cranes, access equipment and site transport.
    - Cleaning and security services.
    - Water and electricity.
    - Stationery, telephones, radios and computers.
    - Provision of insurances and bonds.

To calculate these costs it's necessary to complete the project schedule and work out the resources required to construct the project. Importantly check subcontractor's quotes to see if they have overhead allowances that must be included into the tender overheads or if there are additional items which the contractor must allow for and which must be added to their overheads.

Once the total costs of the overheads have been calculated it's important to decide how they will be claimed in the tender. Some contractors elect to spread the overheads proportionally through the various bill items and add this to each item. This can have adverse consequences if the quantities decrease, which would then result in the contractor being paid a lower value for their overheads, resulting in them recovering insufficient money to cover the actual costs incurred. Of course if the opposite happened and the quantities went up the contractor would recover more overheads than was envisaged and required, which would add to their profit.

I prefer to show the overheads as separate line items in the bill of quantities so there's less risk of being underpaid. It's also more transparent if a variation is lodged for additional time and overheads when the project duration is extended.

When showing the overheads separately they can be broken down into:

- fixed overheads which are normally the costs to establish and set up the facilities, mobilise people and purchase insurances and bonds as well as the demobilisation costs
- time related overheads which are related to the monthly running costs of the project
- value related overheads which are directly influenced by the value of the work completed

It's important to nominate in the tender how the overheads should be paid. Many of the overhead costs are incurred near the start of the project. However, if the client pays the overheads in proportion to the elapsed time on the project, or in proportion to the value of the works completed, then the contractor could experience a prolonged period of negative cash flow.

At times, to improve the cash flow, you could elect to place a higher value into the fixed overheads which are paid near the beginning of the contract, and a reduced value in the time related portion. There is then a risk that if there's an extension of time variation the contractor will be basing their claim value on these reduced time related overhead costs which will result in them earning an amount less than they actually require.

Insurances

Check what insurances are required for the project. In some cases the client may have insurances in place. Read their documentation to ensure they adequately cover your work. It may be necessary to purchase additional insurance to cover events which the client's insurance doesn't cover. Often the client's insurance may also have deductibles or excesses which are too large for the contractor, so again the contractor may have to purchase additional insurance.

It may be necessary to seek expert advice since some insurance policies can be quite complicated.

On large projects the costs of insurance can be quite large, particularly if the insurer perceives there to be additional risks.

Sureties and bonds

The contractor must understand what sureties and bonds they must provide, and ensure they will be able to obtain them should they be awarded the project. Failure to provide them at the start of the project could result in the project being terminated which will be both embarrassing and costly.

The cost of providing the sureties and bonds should be allowed for in the tender price.

Subcontractors

Sometimes the client elects to appoint some of the subcontractors who are termed as nominated subcontractors. These subcontractors will perform work on the project and the contractor is expected to control and monitor them. When the tender is issued the client may already have chosen these subcontractors and they'll be specified in the documentation. However, often the client hasn't selected them and will only indicate the work which is to be performed by a nominated subcontractor. Sometimes they will include a provisional amount for these works which the tenderer must include in their price.

It's important the tenderer understands the risks of accepting the client's nominated subcontractors since they will be responsible for their performance. Where possible, the tenderer should try and influence their selection. If the subcontractors are known at tender stage the contractor should investigate them to better understand how easy or difficult the subcontractor is to manage, and price their tender accordingly or qualify it appropriately.

Some of the work may be done by subcontractors appointed by the tenderer and these subcontractors must receive a request for a quotation as soon as possible so they have sufficient time to provide a meaningful price.

The request for quotation must include:

- the scope of works
- the date when the price must be received
- tender drawings
- special project conditions
- specifications
- the schedule relevant to the subcontract scope of work
- the schedule for the overall project
- the conditions of contract

When the subcontractors' prices are received the quotes must be carefully analysed to check:

- they've priced the scope fully
- they understood the project conditions
- they haven't excluded anything
- what they expect the main contractor to supply (such as scaffolding, cranes, offices, storage and accommodation)
- specific qualifications that could affect the cost of works being priced

Subcontractors' tenders have to be compared to other prices received for the same work. The adjudication can be complicated because often the total prices can be similar but within the different quotes there can be wide variations of prices between individual items, as well as differences between products offered, and the terms and conditions of the subcontract tender.

The contractor must allow in their price for the management and supervision of the subcontractor, as well as the services, facilities and equipment which the subcontractor may have excluded from their price and which the contractor must supply.

Provisional sums

Sometimes clients aren't sure what they want in an area, or they have insufficient information for the contractor to price a section of works. The client may then provide a provisional sum in the tender document which must be added into the tender price. I've heard of several occasions where contractors forgot to add these provisional sums to their tender sum, resulting in them submitting a price which was too low.

The client should state if these provisional sums include or exclude the tenderer's mark-up. In many cases the tenderer may have to allow a separate amount for their profit on this work which should also include their overheads, management, supervision, equipment, services and facilities required for the work.

Allowing for cost increases

It's one thing to work out the cost of doing the work now, but what happens when prices increase after the tender is submitted? We know wages increase annually, fuel prices change monthly, or in some cases even daily, and suppliers frequently increase the price of materials. Of course this usually isn't a problem for a project which is only a month or two in duration and starts soon after the tender is submitted because the increases over this short period should be relatively small.

However, many larger projects start several months after the tender is submitted, and they're often a year, or even several years in duration. Over this period there may be several wage increases and the price of petroleum and other materials could increase significantly. It's necessary to understand these potential increases, which could be significant in countries with a high rate of inflation or with a volatile exchange rate, and make allowances for them.

Naturally the allowances will be guestimates since the increases are usually unknown, although sometimes with wages there are agreements locked in for a period of several years. Other items, like petroleum, have volatile prices varying according to supply and demand, foreign exchange rates and world events.

Unfortunately contractors cannot always allow for the worst case scenario of the maximum potential increases because then their tender will become uncompetitive. It's therefore necessary to take a realistic view of what increases can be expected using historical data as a guide.

To work out the allowance for these cost increases, break the tender down into the different cost components that may be affected by price increases, such as the quantity of petroleum, labour, steel, and so on. Work out the average increase expected for the item, which depends on when the item will be used according to the tender schedule. Apply this increase to the value of the item to calculate the additional cost which should be allowed.

Sometimes it's possible to mitigate against some or all of the risk, by either asking suppliers to provide a price fixed for the duration of the contract, by procuring the materials near the start of the project and storing them on or off site, or by asking the client to absorb some of the risk by paying escalation, or rise-and-fall on the contract.

Escalation and rise-and-fall

Sometimes clients allow contractors to ask for escalation, or rise-and-fall, on the contract price. This is particular common in countries with high inflation.

Escalation is a factor that's added onto the contract price to compensate the contractor for price increases that occurred during the contract period. It's calculated using the contract value and the assumed percentages of the various components, multiplied by the factor by which each item increased as the work proceeds. Escalation factors are usually calculated and published by the Department of Statistics or some similar government body in the country where the project is being undertaken.

Rise-and-fall is similar, and the contractor normally nominates in their tender which items in their contract will increase in cost. Each item would require an estimate of the quantity to be used as well as its starting cost. In the course of the project the contractor claims, and is reimbursed, for the difference in the final cost of the item relative to the cost at tender stage.

Many clients may accept escalation, or rise-and-fall, although generally they prefer fixed price contracts where the contractor has to allow for the risks of price increases.

When clients adjudicate tenders, comparing a fixed price tender to one with escalation, they'll often apply their estimate of the increases and come up with their cost of how much the escalation will add to the price of the tender. The danger here is that clients will often be conservative and add a bigger value than necessary, which could make the tender with escalation more expensive than the one which is fixed.

There are other risks with these formulas in that they don't always compensate the full portion of the increases. In addition, sometimes prices go down and the contract sum will then be adjusted downwards.

Profit

The amount of profit or margin on a project will depend on numerous factors:

- The contractor's policy governing the profit they add to their tenders.
- How important it is for the contractor to win the tender. If a company needs the work the profit level could be reduced to improve its chances of winning the tender, while if they aren't desperate for the work they could add more profit.
- The size and type of project. Contractors who subcontract portions of the work may have a reduced margin on the subcontracted portions. Contractors may apply a lower profit to larger projects and a higher margin to smaller ones.
- The complexity and risks of the project. High risk complex projects may have a higher profit to compensate the contractor should problems be encountered on the project.
- The client and location of the project. Contractors may add a higher profit to compensate for the risks of working for certain clients or for undertaking projects in remote locations.
- Other bidders on the project. If it's known who they are, and if their numbers are limited, assumptions may be made as to the margin these competitors will apply to their tenders. The margins of competitors could be influenced by the amount of work they have on hand and the type of project they prefer to undertake. A rough idea of their competitiveness can be gauged by the prices they submitted for recent projects. Of course this isn't an exact science, but is a guide to the profit which can be applied.
- Knowing the client's budget also gives an indication of the tender price the client is expecting and in some cases it's possible to adjust the profit

to ensure the tender comes in below their budget. Of course sometimes the client's budget isn't correct and it shouldn't be the only determining factor of the contractor's price.

- The accuracy of the tender. When the contractor is confident their price is correct they can apply a smaller profit, while a less accurate price may require a higher profit to compensate for any possible deficiencies and errors in the tender calculations. (However if you have already added a contingency to cover the company for any errors in the price it shouldn't necessary to increase the profit as well.)

The level of profit is normally a decision taken by senior management within the company.

Once the level of the profit is decided the next decision is where the profit should be added. This can sometimes be complex and needs carefully consideration. The simplest is to apply the profit uniformly across all items in the schedule of quantities.

In some cases when the contract is re-measurable the contractor can compare the estimated quantities the client has provided with those they expect to encounter on the project. Often there are differences. Obviously, if the final project quantities end up being lower than the tendered quantities and profit has been applied to these items, then the contractor will earn a smaller profit than estimated. The reverse happens if the quantity increases. I've often reduced the profit on items I guessed would reduce in quantity and applied a higher one to items I expected would increase in quantity. This can be risky if these assumptions are incorrect.

Always seek ways to maximise the tender profit by thinking up and providing alternative solutions, methods, schedules and materials.

Alternative tenders and other incentives

In most cases the client presents a design and construction methodology for the project which tenderers are expected to price. Nevertheless, contractors may propose alternative designs, methodologies or materials which they consider to be cheaper options. Most tenders specify that the contractor must price the original option (a conforming bid), even if they intend to price alternatives.

I've on a few occasions won tenders based on alternate solutions. Often we only passed on some of the savings created by the alternate solution, and have taken the remaining savings as additional profit. If the client has to make modifications to their design the savings offered should be sufficient to make the offer attractive, and also pay the additional costs the client incurs in modifying the design.

Contractors could also offer alternative materials which are more readily available or cheaper than those specified. The tenderer will need to convince the client of their suitability.

There are other tactics to make the bid more attractive. These could include finance terms, early payment discounts, or even offering alternative schedules.

Cash flow

It is imperative that contractors understand the cash flow of a project. Many contractors fail, not because they are poor contractors, or because they're losing money, but because of inadequate cash flow. Many of the cash flow problems can be attributed to contractors accepting unsuitable payment terms, or carrying out projects which are too large for them.

In addition to the contract payment terms and conditions, the contractor ought to consider the payment terms they have with their subcontractors and suppliers. In some cases a supplier may have to be paid a deposit to secure the purchase of a major item of equipment. Most clients don't pay for unfixed materials, even though they may be on the project site.

It's therefore vital in the tender process to prepare an accurate cash flow forecast. This should take into account when the client pays the contractor and the portion withheld for retention. Compare this with when payments will be made to suppliers, subcontractors, personnel, and for other services. Unfortunately most projects have a negative cash flow until they're almost complete, so the contractor must ensure they can sustain this negative cash flow.

Some ways the Estimator can improve the projects cash flow may include:

- requesting an advance payment from the client
- requesting the client pay the monthly valuations earlier
- requesting the client reduce the retention money, or replace it with a guarantee
- front loading the tender (which is where the work done near the beginning of the contract has a higher value, or a higher profit than the balance of the works)
- setting milestone targets when the contractor will be paid a percentage of the contract sum (this is often used by contractors building houses who demand a certain percentage of the total project price be paid when the foundations are complete, a further percentage when the walls reach roof height and so on) and normally these payments are set at a value higher than the actual value of the works completed, so many clients are reluctant to accept these terms as it exposes them to risk should the contractor fail to complete the project
- negotiating to pay subcontractors and suppliers later
- holding cash retention from subcontractors
- requesting the client pay for unfixed materials

Cash flow can be adversely affected by clients who pay valuations late. Consequently, it's important to consider the client's ability to pay progress valuations in full and on time.

Some clients request the contractor to include a valuation schedule in their tender which is used by the client to work out their own anticipated cash flow. It's important that this schedule (or client's cash flow projection), is accurate so the client can correctly forecast their financial requirements. If it isn't accurate the client may not have sufficient funds available to meet particular monthly payment claims from the contractor.

Day-works and supplementary rates

Most tenders request the contractor to complete a schedule of day-works rates. Usually the preamble to this section specifies what the rates should include. For instance, wage rates would normally include the basic wage rate plus allowances, holiday and sick leave pay, leave provisions, bonuses, protective clothing, accommodation, transport and small tools. However, in addition many contracts specify that supervision and other overheads should be included as well, which is usually difficult to assess and depends on the type and size of work to be done using these rates. Therefore, where possible, exclude supervision from the labour rate and nominate a separate rate for this.

Ensure plant and equipment rates include everything that's requested, such as the basic hire costs, maintenance costs (including the provision of service and refuelling vehicles, mechanics, oils, grease, cutting edges and spare parts), fuel and in most cases the operator's wages (again the wages would take into account all of the costs mentioned above). In some cases the contract specifies that the rate should also allow for mobilising and demobilising the equipment. This is typically difficult to do, so where possible nominate to exclude these costs.

Most contracts normally ask for a mark-up or profit percentage to be added to the actual cost of the materials that are used in day-works events. If certain materials are specified it's important to allow for the actual cost of the material plus transport, storage, offloading, handling as well as any wastage and bulking factors.

When making use of subcontractors, make sure that their day-works rates are included in the rates submitted in the tender. Often the subcontractor's rates are higher than the rates agreed with the client which results in the contractor losing money if the subcontractor does work on day-works.

Weather

It's essential to understand and consider the likely weather conditions on the project site during the construction period. Rain, winds and periods of extreme temperatures can have a major impact on the productivity of personnel and equipment which will influence the project schedule and the cost of construction. If necessary, additional time may have to be factored into the tender schedule for these possible delays, and additional money added to take into account the cost of these delays as well as the cost of the reduced productivities.

Cognisance should also be taken of the potential consequence of any client delays in awarding the contract or access to the project site. These delays could push the contract into a season of unfavourable weather which may not have been allowed for in the tender. If there are concerns about these potential impacts they should be highlighted in the tender submission to ensure that if there are delays the contractor will be suitably recompensed for the changed conditions.

Risks and opportunities

Most tenders involve some risk. Risks can be dealt with in a number of different ways, for instance the contractor could:

- eliminate the risk by engineering it away, for example using alternative construction methods, which may add additional costs
- assume the worst case scenario and price the full cost of the risk occurring
- ignore the risk and hope it doesn't eventuate
- qualify the risk out, transferring it to the client

In the ideal world the contractor would like to eliminate all risks, but this is obviously not always possible since the additional money added to compensate for the risk will make the tender uncompetitive. It's also usually difficult to pass all the risk to the client as typically they won't accept this. Therefore the answer is usually a combination of the four options above.

While preparing the tender the Estimators should note the risks in a schedule, including how they've been treated. When the tender is finalised the risk schedule should be reviewed to check the risks have been treated correctly and accounted for in the final tender amount.

Some typical examples of risks to consider may be:

- adverse weather
- payment terms including the risk of non-payment
- contractual conditions that place more risk with the contractor, or conditions which are ambiguous
- industrial unrest
- possible shortages of critical materials
- the availability of equipment and people
- specific engineering problems
- the project schedule
- high penalties or liquidated damages
- poor ground conditions
- lack of information
- political instability

Arrange the risks into the following categories; financial, commercial and engineering.

Against each risk allocate the likely chance of the risk materialising as a percentage (with one hundred per cent being a risk that will definitely occur). Allocate a cost to the risk should it eventuate.

When preparing the risk schedule it's important that you don't see ghosts, adding risks that are unlikely to occur or may in fact be non-existent.

Just as there are risks, so too are there opportunities. These opportunities may be:

- the project scope increases which may give the contractor better utilisation of their resources and increase the project's profit
- further work opportunities

- the client is disorganised which could provide opportunities for claims and variations
- the possibility to re-engineer the project making it easier to construct

List the opportunities in an opportunities schedule with the possible positive cost impacts for the project. Allocate a percentage to the opportunity in accordance with the likelihood of the opportunity materialising.

When assessing the additional costs to be added to the tender for possible risks, also consider the potential additional profit from opportunities which the project may have, and try and balance them out.

Bid (or tender) bonds

Some tenders require the contractor to submit a bid bond with the tender which is normally equal to 10% of the tender value. Failure to submit the bond will invalidate the tender. These bonds could take several weeks to arrange so they need to be requested from the bank at the start of the tender process. Often there is prescribed wording for the bond.

Because the bond is dependent on the tender value it's important to do an accurate estimate of the value as soon as possible and update the bank on any significant changes as they occur.

Form of offer

Most tenders require the tenderer to fill in a standard form of offer which is included with the tender. I've heard of contractors forgetting to complete the form, or filling in the incorrect price. The form of tender is a legally binding document and the client is entitled to accept the value entered on it.

Tender covering letter

It's good practice to include a tender covering letter which should:

- thank the client for the opportunity to tender
- give a brief overview of the tender submission
- reference any qualifications
- amplify why the contractor is best suited to undertake the project
- place emphasis on attending to the client's concerns and priorities

The letter should be brief and pertinent and no more than a couple of pages in length.

Tender qualifications and clarifications

Most contractors will have some qualifications for their tender. These would include the conditions they are willing to accept in order to undertake the works. Furthermore, the contractor may want to clarify what they've actually priced. The shoddier the tender details and information provided by the client, the more items have to be clarified. The contractor must ensure that what they have priced has been fully described in the tender documents and if not, it needs to be clarified.

Additionally, I've sometimes highlighted in the clarifications an item I've priced which I believe other tenderers may have missed. This brings it to the client's attention so they will either ask you to remove the item from your price (if it's unnecessary), or they will request the other tenderers to ensure they've priced for it (which may mean their prices increase).

Typical exclusions and clarifications may include:

- unforseen or unknown ground conditions
- dealing with artefacts and hazardous materials
- unforseen or unseasonal weather conditions
- the correctness of existing structures and ground elevations
- the location and quantities of the client provided facilities
- the tender schedule
- location of existing services
- payment terms and conditions
- capping of penalties and liquidated damages, and their rate of application
- alternative materials or designs
- office and laydown requirements
- dates by when information, drawings and client supplied materials are required
- the provision of a payment bond in lieu of retention money
- assumed working hours
- the tender validity period

I've known some contractors to have pages of qualifications in their tender, some being fairly minor items. I believe this practice could jeopardise acceptance of their tender, so some care needs to be taken to ensure only those conditions which are really important are included.

Checking

It's important to check the final price to ensure:

- it includes all the client's requirements such as:
  - the provision of offices, facilities, services and security
  - their site restrictions including working hours and mobilisation and site induction procedures
  - quality procedures and tests
  - staffing
  - deliverables
  - special requirements
  - spare parts
  - insurances
  - warranties
  - restrictions with tying into existing services
- the calculations are correct
- all items have been priced and included in the final tender price
- it includes all provisional sums
- it has allowed for client supplied items and materials
- it allows for commissioning, including the power, water and chemicals

that may be required for this process
- it includes for all taxes and duties

Tender submission

Don't underestimate how important the tender submission is. This must be as professional as possible. In the past, I've often attended tender clarification meetings where the client has complimented us on our tender submission and they have possibly been swayed to award the project to us.

Some or all of the following could be included in the tender submission:

- an index
- the transmittal or covering letter
- the final price or form of tender
- commercial and technical clarifications and qualifications
- the tender schedule
- a break-down of the price
- the contractor's proposed project management organisation chart and curriculum vitae of senior staff
- a list of equipment and subcontractors
- the deliverables the client has requested (which may include proof of insurances, cash flow and histograms)
- day-works rates
- company profile including a list of similar projects the contractor has completed
- safety information and documentation
- quality documentation (keep the information relevant to the project and demonstrate the company has a clear understanding of the client's requirements and will meet and even exceed these)
- environmental management information and documentation
- if necessary, traffic management plans
- the project approach, work methodologies and considerations taken into account in the tender
- company brochures (which may include financial statements and safety statistics)

Include as much information as possible to demonstrate that the company has an understanding of the project, has thought through the construction process, and has the personnel and resources to undertake the project successfully.

Many tenders are scored not just on price but are affected by other items in the submission such as quality documentation, safety plans, methodology and schedule.

Check the submission to make certain:

- all the documentation requested by the client has been completed and is included
- that all the pages have been printed and are included (sometimes in the binding process pages get inadvertently left out, or during printing, errors occur resulting in blank pages)

- the documentation is presented in a logical and easy to read format (The specific documents the client has requested should be highlighted and easy to find. I've on occasion had clients ask for documentation which we had included in our submission but which they couldn't find. This is frustrating for the person adjudicating the tender and can lead to them assuming it isn't available, resulting in the tender being disqualified.)

Ensure sections such as safety and quality are included in their own separate divisions and aren't spread across other areas of the submission. This could result in the reviewer only receiving, and reviewing, a portion of the relevant information.

Remember, tenders are often reviewed by a number of different people, or departments, within the client's organisation. The tender is frequently split into the relevant sections which are then circulated to various people who aren't even directly associated with the project but will, for instance, be requested to review all the tenderers' safety documentation. If they don't receive the full submission from a tenderer they'll assume it wasn't included and will give the tender a poor rating which may adversely affect the contractor's chances of being awarded the project.

Joint ventures

Joint ventures are normally formed by two or more contractors for the purpose of sharing risk, to pool resources for a large project that they independently couldn't perform on their own, or where one of the partners has an expertise in an area the other partner doesn't, (for instance one of the partners may do the mechanical works and the other the civil works).

Before tendering in a joint venture ensure that the client will accept a tender from the joint venture.

Case study:

I worked for a contractor that submitted a tender worth several hundred million dollars in joint venture with another contractor. However, neither company had asked for the client's approval to submit the tender as a joint venture. Consequently the bid wasn't accepted and we had wasted considerable time and effort.

The letter requesting permission should highlight the reason for forming the joint venture and what benefits it will bring to the client and their project.

Most clients will require that the joint venture partners be jointly and severally liable for executing the project.

Also, before submitting the tender the joint venture parties should have a signed heads-of-agreement, spelling out the terms of the joint venture.

The parties should agree:

- the percentages each will have in the joint venture
- who the lead partner will be
- the resources each party will contribute

- the name of the joint venture
- the address for the joint venture
- who will pay the costs of the tender
- how the tender will be done and who will be responsible for the submission
- a schedule (program) for preparing the tender
- the rates for the resources each party will contribute to the project

Before the tender process starts there should be agreement on the rates which will be used for pricing the works and what these include. For example, the rates for the different levels of staff and whether these rates include for all costs like sick and leave pay, as well as bonuses, and whether the equipment costs include for spares and maintenance and if there are minimum hours the items should be paid for.

Summary

- Tenders should be compiled by people who are familiar with the company's resources and capabilities and who understand the construction process, costs and methodologies.
- Tender documentation should be checked to ensure all of the sections and pages have been received.
- The document must be carefully read and understood and any discrepancies or problem clauses should be noted.
- The contractor should visit the site and note anything which may impact on the construction of the project.
- It's important to prepare and resource a tender schedule.
- If required, work out the quantities for each task or item.
- Get quotes for materials and for works which will be subcontracted.
- All items, or tasks, should be priced from first principles and should allow for the conditions that are relevant to the project and its location.
- Price the overheads needed to do the works and add in any company overheads which need to be included.
- Decide on an appropriate and relevant profit margin and where it will be added.
- Check the project's cash flow to ensure that the contractor has suitable financial means to cover periods when it will be negative.
- Prepare a schedule of risks and opportunities and if necessary take some of these into account in the price, or qualify them out.
- Submit a schedule of qualifications and clarifications which outlines what the tender price includes, what has been excluded, and what has been assumed.
- If possible, provide alternative prices or methods of construction which could be attractive to the client.
- Ensure that the tender submission is complete and includes all the documentation the client requested.

- Check the tender calculations and ensure that the correct figures have been entered into the submission.
- Ensure that the tender has allowed for the likely project weather conditions.
- Obtain the client's permission before submitting a tender as a joint venture.

Chapter 4 – Securing the Project

It's one thing to submit a tender, but even if the price is the lowest the chances of being awarded the project are often less than 50%. In fact, it's often preferably to be awarded a project from second or third place since it means you are getting a higher price. For many clients it's not only about price – they want a contractor who produces a quality product, on time, without any safety or environmental incidents, with minimal fuss, no industrial relations problems and without unwarranted additional claims. Experienced clients understand the additional premium to engage a good contractor more than compensates for delays or problems that may occur on their project when a cheaper contractor is used. Though, unfortunately, some clients are only focussed on getting the project as cheaply as possible – sometimes it is best not to work for them.

It's important to make the bid as attractive as possible, sell the company's abilities to the client and demonstrate you're professional. This is done by submitting a complete and well-presented tender. In addition, respond in full and in a timely manner to all post-tender correspondence, and be totally prepared and responsive at post-tender meetings.

Open the door – price

Price is important; no client is going to talk to a contractor whose tender price is a lot more expensive than that of the other bidders, or whose price is over their budget. As discussed in the previous chapter it's important to have the right price, which not only ensures the contractor shouldn't lose money on the project but is also one that's competitive when compared with the other bidders' prices.

Sometimes Estimators can over think and overcomplicate a project, pricing things that may not become a reality, making their price too expensive. During the tender process it's important to carefully weigh up the likely occurrence of an event, and in some cases rather than pricing for the event the contractor can choose to accept the risk of it occurring or exclude it by qualifying their tender accordingly (Refer to Chapter 3 for more details on assessing and dealing with risks in tenders). If the tender price is of interest to the client they will normally invite the contractor to a tender clarification meeting where the qualification is likely to be discussed. The client may accept the qualification or they may discuss the potential issue, often clarifying it for the contractor. If it's not as serious as first thought the contractor could withdraw their qualification without additional costs. In the worst case scenario the contractor may have to withdraw the qualification and add additional money to compensate for the event. However, at least the contractor has had the opportunity to meet the client and promote themselves.

This doesn't mean that the tender price should be absurdly low with numerous qualifications, because very few clients will find this acceptable. (Note that some client bodies will not accept tenders that contain qualifications and a non-conforming bid, listing savings or proposed alternative materials may have to be submitted together with an unqualified conforming bid). The contractor must take on some risk, and the price should be a reasonable assessment of the work, taking into account reasonable risks.

Discounts and savings

To reduce the price further, making the bid more attractive, it's possible to offer the client discounts and savings if they:

- pay the monthly valuations earlier
- reduce the amount of retention money held, or release it sooner
- allow the contractor to replace the retention money with a surety bond
- have specified the surety must be provided by a bank, allow the contractor to provide a surety from an insurer rather than a bank (which is often cheaper for the contractor)

Sometimes clients break down their projects into smaller separate tenders, but if the contractor is interested in constructing a few of them there may be an opportunity to offer a discount if the client awards two or more projects to the contractor. There are normally savings to the contractor in using shared resources, the benefits of which can be passed on to the client. In addition, the advantage of having one large project made up of smaller projects instead of one small one, may be attractive enough that the profit can be reduced and the saving passed on to the client as a discount.

Understanding the client's needs and priorities

By understanding the client's priorities the contractor can ensure their tender takes these into account. The tender submission should emphasise that the contractor can achieve these criteria and priorities. For example, if the client is concerned about safety then the contractor may need to allow for additional safety personnel in the project price and as part of the presentation the tenderer could emphasise their safety achievements on past projects as well as outline the steps they will implement to ensure the project is constructed safely.

A tender that addresses the client's needs and concerns may be favoured over another tender, even if that one is cheaper.

The client's priorities are often emphasised in the documents or discussed at the tender site inspections. Also, if the contractor has a relationship with one of the client's personnel it's possible to find out what their priorities are.

Some concerns the client may have are:

- achieving the milestone dates
- producing the correct quality
- industrial relations harmony
- the contractor's ability to work around the client's activities
- the contractor's expertise and knowledge

- that the contractor has sufficient resources
- the contractor's ability to work with other contractors in restricted areas
- that the contractor understands the project and its complexities

Case study:

We priced one project in joint venture with another contractor as the project was too large for us to undertake on our own. During the tender process we heard the client didn't like working with joint ventures so in our tender submission we tried to address all of the client's concerns regarding joint ventures and emphasise the advantages our joint venture would bring to the project.

Often clients have completion dates and milestones that are difficult or impossible to achieve. However, it's not always the final completion that's important to the client, but rather for them to get beneficial access to parts of the facility during the construction phase to enable them to start installing their equipment. If the tenderer understands what the access requirements are, it could be possible to give the client their access on the required dates without committing to what might be an impossible task of handing over the complete facility as one milestone.

The tender scoring and adjudication process

Often the client has a defined method they will use to score and adjudicate the tenders which usually takes into account a number of factors including the price. This process often follows a set formula which is outlined in the tender document. The contractor must understand how this formula works, ensuring they maximise their scores. The items the client will look at should be clearly highlighted and must comply with the client's requirements.

For instance, some clients may score the contractor's price as 90% of the overall score with the remaining 10%, say, being made up of various factors such as the local ownership of the company, the number of indigenous employees, the amount of money that will be spent in the local community, the experience of the proposed construction team, the resources available for the project and the contractor's safety record.

The contractor shouldn't overstate what they will achieve because there are often penalties that will apply should they later fail to meet their commitments.

Often the contractor can improve their scores for some items by committing additional money to them. However, the extra costs might eventually outweigh any advantages gained from maximising the scores. For example; using local suppliers may increase the costs, inflating the tender price, or the greater use of local materials could improve the tender score, but their cost could increase the tender price and outweigh this benefit.

Sometimes it's possible to come up with alternative solutions and ways to improve the way the tender will be scored. It's worth spending time and effort in ensuring the tender will receive the highest realistic scoring.

It's also advantageous to consider how competitors will be scored since this could influence both your price and how you decide to best maximise your scoring opportunities.

Differentiating the company

I've discussed the importance of a well presented tender submission, as well as portraying a knowledgeable and professional image at post-tender meetings and negotiations. To win the project it's imperative that the contractor is able to sell the company, differentiating it from other contractors, and convincing the client they are the best contractor for the project. To do this the contractor should emphasise their:

- reputation of successfully delivering similar projects
- experience and knowledge
- availability of resources
- suitable personnel, highlighting their experience with similar projects and their tenure with the company
- reliable subcontractors and suppliers
- excellent safety record
- good quality work on other projects as well as their quality systems and procedures
- understanding of the project
- commitment to meeting the client's needs
- financial capacity
- proactive approach to avoiding and solving problems
- ability to work with the client and their team
- good environmental management record (their 'green credentials')

Do not talk badly of competitors unnecessarily, but do mention items or considerations which you may have thought about, but which other contractors might not have.

Post-tender correspondence

After the tender is submitted there's usually correspondence from the client which requires a response – often immediately. It is essential the tender submission clearly indicates the name and contact details of the contractor's representative who the client can contact (particularly when the tender is submitted just before a holiday period or when the Estimator is going on leave). It's embarrassing, and could cost the contractor a potential project, if the client sends correspondence which isn't replied to because the person it's addressed to is unavailable.

Correspondence from the client should go to the designated contractor's representative, who should coordinate the answers from the relevant experts within the company, and then respond appropriately to the client. If the client has provided insufficient time to gather all the required information for a meaningful answer it will be necessary to request more time. If the client is unable or unwilling to grant an extension the contractor may have to provide their best estimate, ensuring that the client is aware it's only an estimate and a more accurate answer will follow.

When the client asks questions relating to the tender ensure answers:

- address these questions
- have taken into account all of the impacts, which may include changes to the schedule, increased costs, additional overheads, delayed access, extra mobilisation costs and impacts on the rest of the project (for example, even what appears to be a small change in specification could have a major impact on the schedule if an item has to now be imported)
- are provided on time
- are unambiguous
- take into account previous responses
- are framed in a positive manner
- highlight reasons for some answers, where necessary
- offer alternative solutions where the contractor is unable to comply with the client's requests
- provide a revised tender price which includes the effects of previous answers
- are sent in a formal letter (if an answer is provided verbally in a meeting it should be followed by a letter) addressed to the client's nominated representative
- refer back to the original tender documents, terms and conditions, and exclusions and qualifications
- are carefully thought through
- including all supporting documents and calculations, are filed in the tender file so that it's possible at a later stage to understand how the revised tender sum was arrived at

The answers will normally form part of the final contract price and document.

Case study:

One of my projects had a financial problem because the person who compiled the estimate was contacted by the client after the tender was submitted and asked to provide a price to supply and install three hundred metres of electric cable in the ground. The Estimator looked at what he considered was a similar item in the tender and calculated that the cost for the additional cable was eighty seven thousand dollars. He submitted the price in an unqualified one line email to the client even though we had no specifications for the cable or its installation.

When we were awarded the project the client insisted we had to do the work for this price even though when we priced the installation accurately it came to over five hundred thousand dollars. Before we even started the contract we were going to lose more than four hundred thousand dollars – all because the Estimator was rushed into giving an answer at the last minute.

(Incidentally, I must also question the client, since it should have been obvious that eighty seven thousand dollars was far too cheap for a project in that particular region!)

Case study:

Carrying on with the example above, compounding our problem further, the Estimator was later asked to revise the project price to include another change. When this revised price was submitted the cost for the additional cable wasn't added into this new tender price. The revised price was accepted by the client and they awarded the contract for this new value assuming it included all their previous questions and price revisions, including the eighty seven thousand dollars to supply and install the cable.

From the above it can be seen that every time a revised price is submitted it should be clear what's included and what's excluded, and should take into account the previous questions and answers.

Unfortunately, often the questions are sent several weeks, or months, after the tender was submitted and the Estimator has moved on to other tenders or projects. It's important the Estimator takes time to analyse and answer the question, referring to, and reading the original tender submission to refresh their memory. Failure to do so can result in costly errors.

Case study:

I worked on one project where the original tender price submitted was nearly twelve million dollars. The tender was awarded for a value of six million eight hundred thousand dollars. Between the tender submission and the final contract award over five million dollars had been cut from the tender price. This was achieved by the client accepting alternative materials and undertaking some of the work themselves. In addition the tender team made savings in their pricing. However, we could only find records of where half of these savings came from. No one could remember how, or why, the balance of savings was given. It was no surprise that we lost two million dollars on the project.

It's essential when subcontractors or suppliers could be affected by these questions that they are consulted to allow them to modify their prices if necessary. Nonetheless, they do need to understand this mustn't be treated as an opportunity to add money to their quote without reason, making the contractor's tender price too expensive.

The fact that the client is talking to the contractor should be kept confidential so the other bidders don't hear of this. If they know the client is talking to you they may come up with a strategy to ensure they are viewed more favourably. Any subcontractors or suppliers involved in these questions should be cautioned to keep them confidential.

Follow-up

If you haven't heard from the client after a week or two it's good practice to contact them to find out the status of your tender and the project. This shows the client you are interested in the project. You may also obtain some inside information which could give you an advantage over other contractors.

If the client doesn't like your price or tender try and find out the reason so that you can improve your processes for the next tender.

The sooner you know your tender was unacceptable the sooner you can try and secure other work

Post-tender meetings with the client

Many clients request the lowest tenderer, or maybe even the lowest three tenderers, to attend a tender clarification meeting. Before the meeting the client will usually send out an agenda.

In preparation for the meeting:

- read through the agenda and ensure you're prepared
- go through the tender again to refresh yourself on the details
- ensure that everyone attending the meeting is familiar with the project and can add value to the discussion (often senior managers or executives from the contractor attend, many of whom are unfamiliar with the tender and the project constraints, leading them to make inappropriate comments, promises and commitments – ensure they are fully briefed and understand the risks and constraints of the project)
- ensure you know the date, time and place of the meeting, as well as the contact person and their details (it's also useful to know the name of the meeting room since I've regularly arrived for a meeting at a large organisation and the receptionist has been unable to find the contact person because they are busy meeting with another contractor)
- take a complete copy of the tender submission, filed correctly, including the calculations and the post-tender correspondence (there's nothing worse than being asked a question in the meeting and having to page through file after file, looking for the relevant document)
- take sufficient business cards, a note pad and pen

The size and the make-up of the contractor's team attending the meeting will depend on a number of factors:

- the size of the project
- the agenda provided by the client (for example, if there's an item on quality it may be prudent for the proposed project Quality Advisor to attend, or, even perhaps the company's Quality Manager)
- who the client's attendees are
- what the client's known priorities are (for instance, if safety is important take the company's Safety Manager as well as the proposed project Safety Advisor)

Of course it goes without saying that everyone should be dressed neatly and arrive for the meeting on time. It's often useful to arrive early as it may well provide an opportunity to see the other contractors being interviewed, or to meet some of the client's team informally before the meeting which could provide some insight into their thoughts regarding the project and your tender submission.

Case study:

We arrived for a tender clarification meeting to be told apologetically that we wouldn't be starting on time as the previous contractor had arrived late for their meeting – not a good start for them! However, five minutes later the other contractor walked out of the meeting looking embarrassed and we were called in for our meeting almost on schedule.

Sometime later when I met the other contractor at an industry event I asked why their meeting was so short and they told me they thought the meeting was for another tender they had submitted to the same client. They arrived with the documentation for that tender and were prepared to discuss that project, so when the client started asking questions their replies related to the wrong project. When the client couldn't understand what they were referring to they eventually figured out the confusion. Obviously the client was less than amused and quickly terminated the meeting. Needless to say they weren't awarded either of the projects.

At the meeting:

- introduce the team, explaining each person's role on the project
- ensure the contractor's team is led by one person who may delegate others in the team to answer particular questions
- take notes at the meeting even if the client confirms they'll be sending out formal minutes
- take time to answer the client's questions; if uncertain ask them if it's possible to provide an answer later
- listen carefully to the client's concerns and try and address these during the meeting as well as in post-meeting correspondence
- at the end of the meeting summarise the questions that require answers, by when they will be answered, and to whom the answer should be directed (ensure there's sufficient time to get the answers to the client)
- appear positive, and if what the client is asking isn't possible explain why and offer alternative solutions
- don't make promises and commitments that cannot be kept
- thank the client for inviting you to the meeting

Case study:

I've won projects by persuading the client during the tender clarification meeting that we were the best contractor to undertake their project. One particular example was a large project, which the client issued as two separate tenders. We really wanted both portions but the client had already decided they were going to award one portion to us and the other portion to a competitor. During the tender clarification meeting I convinced them that if they awarded the entire project to us it would be a large very important project for us and we would commit our very best team to the project and treat it with the importance that it deserved. This was obviously a risky proposal as the client may have decided that if we weren't going to commit our best team to the project if we were only awarded one portion, then maybe they shouldn't award it to us at all. However I was able to provide such a compelling argument that the client awarded us both tenders.

Some months later our competitor asked me how we had managed to be awarded the complete project since they had been convinced they were going to be given the one portion.

I'm sure I've probably also lost some projects at the tender clarification meetings because of mistakes made or the incorrect answers provided.

After the meeting on returning to the office:

- allocate the questions to the relevant staff and then coordinate the answers into one reply to the client
- when the minutes of the tender meeting are received check that they are correct and are a fair reflection of what was discussed and agreed at the meeting (if there is anything that you disagree with, address this in a formal letter to the client)
- if necessary send a letter after the meeting confirming the discussions and any revisions to the tender offer made during the meeting
- ensure the client's questions are replied to within the agreed time
- send a letter thanking the client for the opportunity to meet with them and offer to answer any further questions they may have

Presentations

As part of the meeting agenda the contractor may have to prepare a presentation. Understand what subjects are required to be covered in the presentation and what the time limit is. If necessary call or email the client for more details.

The presentation should preferably be in PowerPoint or similar. Ensure the client will have suitable equipment or take your own, ensuring there's someone who can operate it.

Presentations should:

- include the name of the company and the names of the people doing the presentation
- thank the client for the opportunity of doing the presentation
- give a brief overview of the company including the company's safety statistics, achievements and policies
- cover the items the client requested
- provide examples of previous experience relevant to the project, including photographs where possible
- demonstrate why the company would be suitable to undertake the project and what strengths the company brings to it
- show the proposed organisational chart for the project with the names of staff and an overview of their roles, qualifications and experience
- exhibit that the contractor has thought through the project and understands how it will be constructed
- briefly run through the project schedule
- highlight any concerns or risks the contractor has
- address any concerns the client may have

- provide a list of proposed major subcontractors and suppliers and outline why they're suitable
- list major equipment that will be used and how it will be sourced
- ask if there are any questions

It's sometimes useful to leave a copy of the presentation with the client.

Tender negotiations

During the various discussions and correspondence the client may ask the contractor:

- to revise particular rates and the overall tender sum
- revise the schedule
- change their construction methodology
- remove some of their tender exclusions or qualifications
- to modify the project scope or specifications and price the changes

The contractor needs to review their tender to assess if the client's requests can be accommodated. In certain cases it may be possible to accommodate some of them. Where they cannot be accommodated, the contractor needs to explain and justify why their tender is correct, and why it cannot be changed. It might also be possible to propose alternative methods or prices to enable the contractor to meet some of the client's requests.

In some cases it's possible to go back to suppliers and subcontractors and ask them to re-assess their quotes to see how they can accommodate the client's requests.

This is all part of the negotiating process and it's important to try and understand the client's final position. Obviously if the contractor has a good relationship with one of the client's team it's often possible to gain inside information as to what the client's real concerns are.

Ensure all copies of letters and calculations relating to these negotiations are placed in the tender file as well as any new information the client may have issued relating to their requests.

A word of warning – don't secure the tender at any cost

Unfortunately some contractors can become obsessed with being awarded a contract. They've gone through the process of preparing the tender, and now the client is asking questions and calling them in for meetings. The negotiations are proceeding well and the contractor can almost smell the success of being awarded the project, but, the client is still requesting a reduction in the tender price or refusing to accept some of the contractor's qualifications. In the heat of the moment it's easy to give in to the client, provide a reduced price and accept a contract which excludes some, or all, of the qualifications. At times, this is done by senior management who don't fully understand the project or their team's concerns, only wanting to return to the office to say they've been awarded a new project thanks to their timely intervention.

I know some contractors are willing to concede almost anything during these tender negotiations to be awarded the project. Sometimes, they even believe they'll be able to find a way during the course of the project to get out of their commitments, or to recover the money they gave away. This often leads to projects that go wrong, lose money and give the contractor a bad reputation.

I'm not saying the tender price is immovable, and nor am I saying the qualifications are cast in stone. However, it's important to do a reality check, work out the minimum price and profit that makes the project worthwhile, and consider the cost and risk of removing the problematic qualifications. After all, I assume the original tender price was arrived at using logic and careful calculation, and the qualifications were inserted for a reason – why then should any of this change?

Sometimes clients pressurise contractors in a meeting to make an immediate decision about something which will affect the outcome of the negotiations. In this case, I suggest the contractor requests a break to confer with colleagues and work through the consequences of the decisions, or preferably, requests twenty-four hours to provide an answer. However, it's important to respond as soon as possible otherwise the client may lose interest and approach another contractor.

Whatever happens with the negotiations, don't make rushed and stupid decisions simply to be awarded the project – decisions which may be regretted later.

Bribery

At no time should the contractor try and influence the client's tender adjudication by offering their representatives bribes, or courting them in any way which may appear as if the contractor is trying to sway the adjudication process. This could result in the contractor being disqualified from the tender process, and in certain cases even result in legal action.

Filing and storage of tender documents

It's essential that the complete tender documentation is filed in an orderly manner. This should include all documentation received from the client, the tender submission, all correspondence, meeting notes and minutes. Documentation shouldn't be removed from the master file.

Subcontractor and supplier quotes, together with all correspondence relating to their quotes, must be filed together with their assessments and adjudication. It should be clear and traceable so project staff can follow the logic of why a particular price was used.

The detailed tender calculations must be retained. These should clearly show what assumptions were made and the equipment, manning, rates and productivities used to calculate the price. They should also reflect changes made during the tender negotiation phase and the basis for these changes.

If the project is awarded to the contractor then a copy of the full set of tender documents and correspondence should be handed over to the project team. The original master copy must be retained, filed and stored safely.

Summary

- Securing the tender isn't only about having the cheapest price, although it's important to come up with innovative ways to make the price attractive to the client, such as:
    - offering discounts
    - proposing alternative solutions
    - offering savings
    - by not overcomplicating the project or pricing all possible risks and eventualities
- It's also usually dependent on:
    - being able to differentiate the company from others
    - demonstrating that the company is the best one for the project
    - taking into account the client's adjudication process
    - satisfying the client's concerns and priorities
- An important part of the tender process is the:
    - post-tender correspondence
    - tender clarification meetings
    - bid presentations
    - tender negotiations
- During these processes the contractor should:
    - be professional
    - ensure answers they provide are clear and unambiguous, taking into account the tender as well as previous correspondence, and are clear as to what has been included and considered in the revised quotation
    - be positive, outlining alternatives or reasons why the contractor may not be able to comply fully with the client's requests
    - carefully think through questions, considering all impacts and implications of any revisions
    - confirm all discussions in formal letters to the client
- Care should be taken in the negotiation process that the contractor doesn't accept conditions or a price which will cause problems during the course of the contract.
- All tender documentation and correspondence must be filed in an orderly way in a tender file.

Chapter 5 – Delivering the Project

In previous chapters we've discussed finding the right projects, ensuring that the tender is compiled accurately considering all costs and risks, and finally that the project has been secured. It's now time to start the project and run it efficiently to ensure that it makes money.

This chapter should be read in conjunction with Chapter 6 which deals with reducing costs and overheads, since much of that chapter is relevant to running the project.

Some readers may work in a small company, being the owner, Estimator and Project Manage. You'll probably think some of the processes outlined below don't apply to you – after all if you prepared the tender why should you have a meeting with yourself to understand the tender process? But it's often just as important to review the tender, even if you did it, before starting the project, because you are probably doing several tenders at any one time so it's easy to forget how you derived the particular price. It's equally as important to follow some of the other steps outlined below, such as planning the project, preparing the schedule, meeting the client and so on.

Plan the project

It's important to plan the project thoroughly before starting work, so the more time available to do this the better. The planning includes:

- thinking through the construction methodology and ensuring the methods selected will be the most efficient and economical
- preparing the project schedule
- taking out adequate insurance
- deciding what work will be self-performed and what will be subcontracted
- ensuring the required resources are available and procured
- ordering long-lead items
- ordering materials which will be required to start the project
- sourcing equipment
- arranging the necessary permits
- preparing and submitting the required paperwork to the client which may include:
  - quality plans
  - safety plans
  - environmental plans
  - traffic management plans

- method statements
- job hazard assessments

Methodology

Careful thought should be given to the methodology of construction. Often the client may have recommended or even specified a methodology, and usually the Estimator would have proposed one in the tender. However, it's often possible to propose other methods of construction which could be more suitable. There's always more than one way of constructing a structure or facility, but some methods will be more efficient and suitable than others, which will vary between projects and locations.

Some factors to consider when preparing the method of construction are:

- the workers' safety
- the facility or structure to be built
- the project schedule
- costs (for example in some areas labour costs are high so it's advisable to reduce the amount of labour required by using more machinery or proposing precast solutions)
- the client's design
- restraints imposed by the client such as:
    - their access requirements
    - the availability of services
    - coordinating with their other contractors
    - access to work areas
    - tie-in and disruption of existing services, processes and traffic
    - restrictions on imports
    - utilisation of local resources
- the availability of:
    - equipment
    - skilled workers
    - staff
    - materials
    - accommodation
    - services such as power and water
- the site conditions such as:
    - topography (for example steeply sloped sites may make it difficult to set-up cranes)
    - ground conditions (for example unstable ground or rock may dictate the rate of progress, the schedule and the type of equipment)
    - access to site (for example the roads may have load limitations which limits the size of equipment or items which can be brought to site)
    - traffic on and around the site which could slow deliveries or limit the hours of work

  - congestion on the site and around the site (for example cranes required to place heavy equipment might not be able to be set-up close to structures)
  - the location of the work area (for example it may be elevated which would restrict access for personnel and materials)
- what methods the contractor's personnel are used to, as well as their level of skill
- expected weather conditions during construction
- the location of the project
- resources available near the project site
- the complexity of the project
- the amount of repetition on the project
- the degree of accuracy required for the finished product
- the finishes required
- the utilisation of resources
- minimising risks

Project schedule

It's important a schedule is prepared which should:

- enable the project to be constructed in the shortest possible time, making efficient use of the available resources, without jeopardising the quality, safety or integrity of the project
- take into account any client imposed restraints, such as; interfacing with other contractors, access dates, working in and around existing facilities and the availability of information
- meet the completion dates that were committed to in the contract (unless the project has changed from the one that was in the tender submission)
- allow sufficient time for planning the project and for mobilisation (on some projects it can take four weeks or more to get personnel through the mobilisation process and on to site)
- adequately show the type of resources required and when they'll be needed
- be approved by the client in writing as soon as possible (without an approved schedule it's difficult for the contractor to claim for variations, late information, late access and extension of time)
- clearly show when access is required to the various work areas
- indicate when information is required (a separate 'information required list' should be prepared which can be updated weekly and discussed with the client at the progress meetings, so the client can be aware of what information is necessary in the next two weeks and notified when information is issued late or when it's inadequate or incomplete for construction purposes)
- be updated regularly (the update must be done correctly, focusing on the critical path activities rather than the overall percentage complete)

- be communicated to the relevant staff so they are aware of the key dates and milestones (Supervisors are often only interested in their section of work and what they need to do in the next couple of weeks, so it's pointless giving them the entire schedule to the end of the project since in many cases it won't be read and will only confuse them – rather give Supervisors a snapshot of the schedule pertinent to their work, even perhaps giving it to them in a pictorial form which can be easily read and displayed on their office wall)
- if needed, be discussed with Supervisors to make sure they understand what needs to be done and why the sequence and resourcing shown is necessary

Don't start before there is a contract in place

Without a contract there's no agreement, no protection for the contractor and no guarantee that the contractor will be paid for the work they do. Therefore ensure that there is a signed agreement in place before starting work.

Furthermore, failure to agree the contract terms and conditions prior to work starting means that when the contractor finally reviews them they have often incurred costs on the project and aren't in a position to persuade the client to alter the terms and conditions to those that are more appropriate or acceptable.

Payment bonds, insurances and guarantees

It's important these are in place before work begins. They are discussed further in Chapter 9

Tender handover

The Project Manager must receive a full set of all the tender documents including:

- the tender submission
- all tender correspondence
- the tender documents, drawings and specifications
- post-tender correspondence
- copies of all subcontractor and supplier quotes used to price the project
- the tender schedule
- tender calculations and price make up, including the calculations for any post-tender revisions

Every project should have a tender handover meeting which is attended by the Estimator, the Project Manager and company support departments as required.

The tender handover meeting should include:

- the Estimator going through the tender submission
- reviewing the risks and opportunities of the project
- a discussion of the tender methodologies
- a discussion of post-tender variations and correspondence

- reviewing the construction strategy
- gaining an understanding of the client's concerns and goals
- informing the company's support departments of what's required from them
- reviewing the urgent deliverables required before work can proceed

Client handover meeting

This is often called the kick-off meeting and is an opportunity for the contractor to meet the client and to understand the rules for the project.

Normally the client would provide an agenda for this meeting. Some items that should be discussed include:

- introduction of both the contractor's and the client's team (their names and responsibilities)
- the procedure for issuing and receiving drawings
- confirming the dates for the submission of project deliverables such as:
    - sureties and bonds
    - proof of insurances
    - the project schedule
    - quality plan
    - safety plan
    - environmental management plan
    - traffic management plan
    - mobilisation schedule
- the requirements for submitting monthly valuations such as when it's required, the format and where it should be sent
- who is authorised to issue and receive instructions and variations
- who correspondence should be addressed to
- availability and location of the services and utilities
- the location of the laydown areas
- permit requirements
- the client's restrictions
- site security and requirements
- date, time and place of project meetings
- quality control procedures
- safety procedures
- confirmation that the project is running according to the tender schedule
- confirmation of the site access

This meeting should be attended by the contractor's Project Manager and members of the project team such as Safety Advisor, Quality Assurance Advisor, Planner and if possible senior management such as Project Director and even on large projects the Estimator responsible for the bid.

Staffing the project

Senior managers should ensure the Project Manager has the support they need. This may include additional training, such as familiarising them with company procedures and policies.

In addition it's advisable for the various departments (such as quality, safety and human resources) to visit the project regularly, spending quality time with the Project Manager, advising them on how to better manage the project. I've often experienced head office staff briefly visiting a project and providing little feedback, support or advice to the Project Manager. However, on returning to Head Office they circulate a report to senior management which is critical of the project's performance. This doesn't help the Project Manager!

Where possible, inexperienced staff on the project should be matched with experienced staff. Recently employed staff should be matched with staff with a longer tenure.

Case study:

One person, who'd previously been a Supervisor, was given his first project in the role of Project Manager. He reported directly to the divisional General Manager who, needless to say, was busy with other projects and securing work. The Project Manager, with no previous experience and little training, was left on his own without help or guidance on how to plan and run the project. To add to the problems, the project wasn't staffed correctly, with only a few inexperienced support staff unfamiliar with the company's procedures. Needless to say the Project Manager soon ran into difficulties, as did the project.

The few times the General Manager visited the project he was critical of how the project was being managed, but offered little help or assistance. Eventually the Project Manager experienced a number of minor personal breakdowns and had to be removed from the project. By this time the project had turned into a mess, with financial and schedule problems, and the Project Manager's career was destroyed before it had even begun.

The first lesson from the above is to ensure a suitable person is placed in the Project Manager position. It cannot just be about filling a position! The person must have suitable experience, knowledge and training for the role. Consideration should be given to the type, complexity and size of the project, as well as the type of client. For instance a Project Manager who has managed large road projects but never a building project probably isn't the best candidate to manage a complex building project.

Secondly, the Project Manager must know what's expected of them. This would include what project reports have to be produced and by when. They must understand what their limits of authority are, the company procedures, and the key performance indicators used to measure their performance.

Furthermore, the project must be staffed correctly. I've often been on projects which were understaffed. In fact I've probably been guilty of having too few staff on many of my projects. This creates stress for the Project Manager and other staff, often resulting in them working excessive hours, sometimes making inappropriate decisions because they've not had time to analyse the facts correctly. Safety and quality are compromised because staff spend inadequate time on site ensuring the work meets the required standards. Details get overlooked. Productivity of the workers is often poor since they are not supervised effectively and don't have the materials and plant to execute the tasks because their managers haven't ordered them.

Of course the converse is also true and the project shouldn't be over-staffed. Too much staff not only leads to inefficiencies and additional costs but can result in the staff becoming bored.

The number and type of staff required for a project depends on a number of factors including:

- the client's specified requirements of the type and number of staff the contractor must provide
- the size of the project
- the complexity of the project
- the size of the client's team
- how good the client's team is and whether they'll be able to deliver quality information on time
- the duration of the project
- whether additional shifts or night work is required
- the skill level of the contractor's workers
- the construction methodology
- whether the contractor will self-perform or subcontract work
- the distance between work areas or the size of the area that the project site covers
- the type of work
- the skills and experience of the contractor's staff

Sometimes the contractor doesn't have to use their most senior or experienced staff on a project; in fact sometimes senior staff can be detrimental on small projects as they:

- become bored and consequently unhappy
- they are often expensive so aren't cost effective
- they may come with a big team
- they are used to working in a big team, with the support of junior staff, so are unable to work on a project where they are expected to do all the work themselves

Some staff are more used to, or suited to, working on their own. When they're placed on a large project working with others and sharing their resources they become unhappy and cause problems.

Staff must:

- be aware of their areas of responsibilities
- know who reports to them and who they report to
- understand what documentation and deliverables they are responsible for and when these must be completed (for example updating schedules, handing over safety documentation and submitting time sheets)
- be aware of their limits of authority and the authorisation processes
- understand the company's systems, standards, policies and procedures

- be competent, with the required knowledge and experience to fulfil the tasks they're required to do
- deal with clients, their representatives, subcontractors and suppliers in a professional manner
- be able to work as a team with the rest of the contractor's staff

The project should have an organisation chart which shows the staff, their positions, responsibilities and reporting structures.

Inductions

All personnel working on the project should attend an induction held by the contractor. This induction should include:

- a welcome by the Project Manager
- an overview of the project
- an overview of the work the contractor is engaged to do
- project progress so far and upcoming milestones
- the project rules
- orientation of the project site including the location of access routes, first-aid facilities, muster points, evacuation routes, eating areas, toilets, and offices
- accommodation arrangements including the accommodation rules
- transport arrangements
- working hours
- disciplinary and grievance procedures
- quality expectations
- safety including:
    - accident and incident reporting procedures
    - personal protective equipment to be used
    - project safety rules
    - particular safety risks and hazards pertaining to the site and the work
    - the location of first aid stations
    - emergency contact numbers
    - the location of muster areas
- environmental including:
    - handling, separating and disposal of waste
    - potential environmental hazards on the project
    - dealing with accidental spills
    - environmentally sensitive areas which should be avoided
    - dealing with the local wildlife on the project
- the site management structure
- the company's values and policies

Many of the industrial relations, safety, environmental and quality issues and incidents on a project can be avoided if all personnel are aware of the applicable procedures, requirements and policies. The time and effort spent on giving a well prepared induction will save unnecessary costs and unhappiness later.

Even when you're the sole person working on the project it's still worthwhile familiarising yourself with the hazards and rules of the work site.

Design and construct projects

Design and construct projects usually require the contractor to manage and coordinate the design process. It's important that the contractor:

- appoints reputable designers and architects who have:
    - the required knowledge and skills to undertake the design
    - design indemnity insurance adequate to cover the contractor should there be an error with the design
    - sufficient resources to deliver the information and drawings in accordance with the schedule
- ensures the design meets the client's requirements in terms of; aesthetics, durability, maintenance and running costs, specifications, output, sustainability and other special requirements including maintaining access where required
- confirms the design complies with the local bylaws, codes and legislation
- checks that the design is suitable for the climatic conditions
- confirms that all permits are in place
- confirms the final product will meet all safety requirements and legislation and will be safe to use and occupy
- ensures the design:
    - is cost effective
    - uses local materials where possible
    - can be built using the skills available
    - utilises the available equipment
    - can be built safely
    - can be built in the time specified by the client
    - does not unduly disrupt traffic or services
    - takes into account the weather during construction
    - allows for the ground and site conditions
    - takes cognisance of access restrictions

The contractor should appoint a suitably qualified and experienced person to manage the design process.

Regular meetings should be held with the design team to review the design and ensure that the information is issued in accordance with the project schedule. It's also necessary to ensure that the client does not add to the project scope, change the parameters of the design, or unnecessarily withhold design and drawing approvals during this period.

Design and construct projects provide the contractor with an opportunity to produce a design which meets the client's requirements but is also constructible and economical to construct using the contractor's resources and expertise.

Site services and facilities

Site facilities such as offices, toilets and lunch rooms must:

- be sited in an area which is conveniently located to the work areas but will not be in the way of future structures
- be safe and structurally sound and weatherproof
- have sufficient lighting and ventilation
- be in good condition and erected neatly, so they portray a good image of the company
- be affordable
- comply with the legislative requirements including satisfying local bylaws and labour agreements
- be large enough to accommodate the expected peak numbers of people, allowing for expansion if necessary
- have their layout approved by the client
- be able to be easily removed at the end of the project
- if necessary, allow for the client's and subcontractor's requirements
- be set up quickly to enable work on the project to commence
- have offices which are comfortable enough for staff to work in, with sufficient furniture and storage space
- have adequate communications
- have secure storage for confidential documentation
- have information technology access points, including computer servers and internet connectivity

Sometimes it's possible to use and modify existing buildings and facilities which could save money and time.

Site facilities should be kept neat, tidy and clean, since this promotes safety, productivity and the image of the company.

Site services such as water and power must:

- allow for peak consumption including all the subcontractors' and commissioning requirements
- comply with legislative requirements
- be planned to not interfere with construction works and be protected from damage
- allow for possible interruptions in supply (where necessary additional storage or backup facilities may have to be installed)
- be metered if the contractor is paying for usage

Subcontractors

Subcontractors can play an important part in the success of the project. It's important that subcontractors aren't chosen purely on price. The subcontractor's ability to deliver the project on time and to the required quality and safety standards is equally important. I've worked on projects where the selection of the cheapest subcontractor ended up costing the project more money than if a more expensive subcontractor had been selected.

Subcontractors are viewed by the client as an extension of the contractor and a failure by the subcontractor can adversely affect the contractor's reputation.

Some important points to note when managing subcontractors are to ensure:

- the contractor's person managing the subcontractor understands:
  - the subcontractor's scope of work
  - who is responsible for supplying what
  - how the subcontractor is reimbursed
- the subcontractor:
  - complies with the safety requirements
  - produces work of acceptable quality
  - works according to the project schedule
- the subcontractor receives access and information on or ahead of schedule and isn't delayed by the contractor or other subcontractors
- regular meetings are held with the subcontractor to discuss safety, quality and environmental matters, as well as progress on the project and any delays and claims, and that minutes of these meetings are distributed to the relevant parties
- subcontractors sign acknowledgement for the receipt of the drawings and information issued to them
- where relevant, the subcontractor supplies shop drawings in accordance with the project schedule, including allowing for obtaining the required approvals from the contractor or the client
- communication with the subcontractor of a contractual nature is in writing (any verbal instructions should be followed up in writing)
- only the contractor's delegated responsible staff communicate with the subcontractor
- action is taken as soon as it appears that the subcontractor could be in trouble
- the subcontractor is forewarned of the contractor's intention to back-charge them for work or services supplied by the contractor and that these charges are invoiced regularly
- the subcontractor is paid in accordance with the contract
- all guarantees and warranties are in place before the final payment is released
- the subcontractor has suitable quality, safety, environmental and industrial relations procedures in place that comply with the project requirements
- subcontractors don't begin work until there's a signed contract in place and they've supplied the required sureties and insurances
- the subcontractor's staff, equipment and their own subcontractors are approved by the contractor
- the subcontractor's personnel attend the contractor's project induction
- correspondence from the subcontractor is promptly dealt with

Understand the contract

Many contractors lose money because their Project Managers have failed to understand the contract or have not acted in accordance with the contract. These failures include amongst others:

- not submitting valuations on time or with the correct supporting documentation
- not ensuring the contractor is paid on time
- carrying out work outside the project scope
- not ensuring the client meets their contractual obligations
- failing to submit claims and variations timeously
- not following the correct insurance claim procedures
- failing to meet the project milestones

It's therefore important the Project Manager carefully reads the contract, noting important items and asking for advice when they are unsure of anything.

Quality control

Rework due to poor quality workmanship and materials causes major additional costs. It's essential that proper systems are put in place to monitor and control quality and that these are implemented from the start of the project.

Refer to Chapter 6

Safety procedures

Projects must be run safely and in compliance with the safety legislation, the client's requirements and the contractor's own safety standards.

Safety must be set up and applied correctly from the start of the project and management and workers must comply with the project safety requirements.

All projects should have:

- safe equipment
- sufficient and appropriate personal protective equipment which must be worn in the work areas
- the correct safety signage
- sufficient first-aid and firefighting equipment
- a sufficient number of people trained in first-aid and personnel able to use the firefighting equipment
- access to a means of communication in the event of an accident
- emergency response procedures in place which personnel are aware of
- emergency contact details readily available
- suitably trained and certified personnel that can safely operate the plant and equipment
- regular tool-box meetings to discuss safety concerns and changes in operational procedures
- procedures in place to prevent alcohol and drug abuse

- proper tagging procedures to ensure the equipment is regularly checked and in working order
- lock-out procedures to prevent the unauthorised use of equipment which isn't safe or that's being worked on
- safety awareness training
- on larger projects, safety committees
- accidents and incidents reported and investigated
- potential hazards communicated to all personnel
- hazardous materials stored in a separate lockable location with the material data sheets accessible
- flammable liquids stored in a well-ventilated store away from flames

Environmental

It's important that projects comply with environmental legislation as well as with the client's environmental plan and permit conditions, along with the contractor's own environmental certifications and policies.

Suitable care must be taken to eliminate:

- dust
- noise
- air pollution
- stormwater run-off
- erosion and silt deposition
- fuel, oil, chemical and other hazardous material spills
- contamination of the ground and surroundings
- the risk of fire
- the spreading of weeds
- waste (which should be recycled where possible)

Care should be taken to protect fauna and flora, including fencing off sensitive areas and ensuring workers stay within the designated work areas.

Drawings

The proper and orderly control of drawings is essential to the success of the project.

- A drawing register should be set-up and maintained. This should be compared to the client's register and any discrepancies reported immediately.
- Drawings issued to subcontractors, suppliers and staff should be issued under cover of a transmission note which should be acknowledged and returned by the recipient.
- A master copy of all drawings must be kept in the project site office. These drawings must:
    - be filed correctly according to drawing number, and if necessary, in their various sections
    - never be removed from the master file unless they are replaced by a revision that supersedes them

  - not be removed from the site office
  - not be defaced or written on
  - be kept up-to-date and superseded drawings should be marked 'superseded' and removed
  - be available on drawing tables where they can be easily referred to
- Drawings must be stamped with their date of receipt.
- Drawings must be issued to the relevant person.
- A master set of all the superseded drawings should be kept in the site office so that drawing changes can be tracked if necessary.
- The Project Manager must:
  - be aware of recently issued drawings
  - be aware of what drawings and information are outstanding
  - ensure drawings are issued to the correct people
  - ensure Supervisors are using the correct drawing revision
- Supervisors should:
  - have access to a clean, dry table on which to lay out their drawings
  - maintain their drawings in a file so they aren't mislaid
  - remove and clearly mark superseded drawings
  - ensure they are working off the latest drawings
  - report any problems or discrepancies with drawings to their Section Manager or Project Manager
- Drawing errors, ambiguities or conflicts must be reported in writing to the client.
- Shop drawings should be:
  - checked by the contractor to ensure they are correct
  - monitored and tracked to ensure they are submitted to the client timeously and that all comments and changes are returned to the originator as speedily as possible

Milestones

It's essential that the contractor meets the project milestones. Failure to do so may result in:

- the contractor incurring penalties
- additional costs to the contractor because they remain on the project longer than anticipated
- the contractor losing credibility with the client
- the client's contractors and work impacting on and restricting the contractor's access

Therefore it's essential that Project Managers and project staff clearly understand what has to be achieved to meet the milestones, and to track progress on the critical path on a daily basis if necessary.

If it appears that progress on the critical path, or in an area that's required to meet a milestone, is slipping then appropriate action may have to be taken including:

- employing additional resources
- moving resources from other less critical areas
- working extended hours on critical activities
- giving priority to critical items in the division of services, equipment and materials on the project
- dedicating key staff to the area

If it appears that the contractor will not meet the date:

- discuss with the client how they can give be given partial access to meet their needs
- see if the client can help in any way
- as a last resort, if it's totally impossible to meet the milestone give the client as much warning as possible (with sufficient notice the client may be able to adjust their dates, reducing the impact and costs caused by the delay)

To meet a milestone the contractor must understand the requirements for the handover, including:

- what work must be completed
- the commissioning requirements
- tie-ins to existing services
- test results (on some projects the facility may have to be completed and commissioned, and then run for several days so that the operations of the facility can be tested before it will be accepted as being complete)
- operational permits and licenses
- operational and maintenance manuals, as well as other documentation specified in the contract
- agreeing and completing punch lists
- training of the client's operations and maintenance staff
- supply of spare parts
- safe access for the client's equipment and contractors
- operation of any services, which may rely on other parts of the facility being completed such as power plants and water treatment plants

Company management must ensure the project staff understand the milestone requirements and are focussed on achieving them and have the required resources.

Daily records, and daily reports

It's important that all projects maintain daily records which should:

- be accurate
- be (preferably) signed by the client's representative
- record the following information:
    - weather conditions

  - numbers and types of personnel and equipment including those provided by subcontractors
  - delays
  - major equipment and material deliveries
  - important tasks completed
- be completed daily

These records could be important should there be a dispute or variation claim.

Meetings

Projects should have regular meetings with the client which are minuted. Project Managers should:

- go to these meetings well prepared with the information requested from the previous meeting, if not already provided previously
- ensure that the minutes are a fair reflection of the meeting and are accurate
- have a list of points that need to be discussed and raise these under the correct sections in the meeting
- take notes of items that need to be actioned
- close out items in the minutes as soon as possible
- immediately on returning to the office action items raised in the meeting
- ensure that the meeting agenda covers items such as; access, information required, outstanding drawings, variations, drawing approvals, delays, progress, other problems or concerns and payments

When possible senior company management should attend client meetings since:

- it's an opportunity to meet the client
- it provides support to the Project Manager
- it affords an opportunity to understand how the project is going and whether the client is happy with the contractor's performance
- they can raise issues which concern them
- clients usually appreciate it when the contractor's senior management take an interest in the project
- it gives the client an opportunity to raise any concerns they may have with the contractor's site management team

Project photographs

Project photographs are useful to:

- record progress, particularly if they have the date and time recorded on them
- show variation and additional work
- record accidents and incidents
- record insurance events
- record equipment or materials that arrived on site damaged or in a poor state of repair

- compile advertisement material
- include in company newsletters so others in the company can see what's involved in the project
- be included in presentations to prospective clients, particularly when trying to secure a tender for a similar project

Permits and documentation

Ensure all permits are in place before work starts and that these are kept up to date. Even if it's the client's responsibility to obtain the permits the contractor should check these are available because the project could be stopped if permits aren't valid, resulting in delays and even causing the contractor to have to demobilise from the project. The client may be unable to obtain the permits which will result in the cancellation of the project, possibly causing financial problems for the client and resulting in the contractor not being reimbursed for their costs.

Permits and documentation vary between regions so it's important that the contractor is aware of the latest requirements and if necessary obtains appropriate advice.

Liaison with the estimating department

Project Managers should regularly provide feedback to their estimating department. This feedback could include:

- the performance of suppliers and subcontractors
- mistakes in the pricing of the project – both positive and negative
- difficulties or advantages of working with the client and their design team
- new methods of construction or new materials
- labour and equipment productivity

In addition it's often useful to invite the Estimators to site so they can witness first-hand how the project is being constructed and compare this with the way it was tendered. In visiting the project the Estimator may even notice items the contractor is undertaking which aren't in accordance with the original tender scope or conditions and which should constitute a variation.

Finishing the project

Many projects are financially successful until the end, when costs suddenly spiral out of control. The main reason for the additional costs is because the project isn't completed on time. When I say completed, I don't just mean handed over, I mean one hundred percent finished. Many Project Managers focus only on handing over the project. However, there's normally more to a project than this. It includes finalising punch lists, finishing and submitting all paperwork (including as-built drawings, quality data packs, guarantees and warranties) and concluding all the contractual obligations (such as commissioning and testing).

To facilitate the timely completion of the project a completion schedule should be prepared near the end of the project. This would include:

- finishing the outstanding items

- commissioning
- connecting to existing services and structures
- completion of the contractor's punch-list items
- final punch-listing by the client
- preparation of hand-over documentation such as quality records, commissioning results, operating manuals and guarantees
- clearing of the temporary site facilities and services

Some of the items which should be considered at the end of the project include:

- obtaining the certificate of practical completion
- getting the release of sureties or bonds and returning them to the institution which issued them
- requesting the release of retention
- putting items of equipment off-hire and transferring them from site
- demobilising all offices and facilities
- reinstating laydown areas and access roads, including obtaining signed acceptance from the client
- agreeing the final accounts with the client
- settling accounts with subcontractors and suppliers
- handing back all accommodation
- disconnecting services
- transferring or terminating personnel
- moving personnel records to the head office
- handing over all spare parts and client-purchased materials to the client
- clearing unused materials
- sorting, filing and archiving project documentation
- disposing of project-purchased assets
- handing over quality documentation, commissioning data, spare parts lists and warranties to the client
- completing the final cost report
- submitting the final project invoice to the client

Summary

To successfully complete a project it's necessary to:

- plan the project including developing a construction methodology which is appropriate, cost efficient, safe and meets the client's requirements and specifications using the available resources
- prepare a schedule and then use this schedule to monitor progress to ensure the work is carried out in the correct sequence and on time
- ensure a contract which adequately protects the contractor is in place before work starts
- confirm all payment bonds, sureties and insurances are in place
- have a formal handover of the tender to the contractor's project team
- hold a handover or kick-off meeting with the client

- staff the project with adequate numbers of suitably trained and experienced staff who understand their duties, the project requirements and the expectations around their performance
- have all personnel on the project attend an induction to inform them of the site rules, safety hazards and project requirements
- if the project includes a design component, manage and coordinate the design process
- establish suitable and sufficient site services and facilities
- manage subcontractors properly to ensure they comply with their contractual obligations
- understand the terms and conditions of the contract
- institute the appropriate quality systems to ensure the quality meets the client's specifications and the contractor's own systems and standards
- ensure all work is safely executed in accordance with safety regulations and legislation, in terms of the client's requirements and meeting the contractor's own safety standards
- ensure compliance with all environmental legislation and permits
- put in place proper drawing and document control systems
- achieve the contractual milestones
- maintain accurate and detailed daily records which should preferably be signed by the client
- have regular meetings with the client which are minuted
- take photographs to record progress, incidents and variation work
- ensure all permits are in place
- provide feedback to the estimating department
- finish the project completely, on time

Chapter 6 – Reducing costs

Let's consider the following example. If a project is valued at eleven million dollars and was tendered at a 10% profit then the project should make a million dollar profit. (Remember the profit is normally added to the cost, so an eleven million dollar project with a 10% profit would have ten million dollars of costs and a million dollars of profit). Say the project actually has costs of twelve million dollars so the contractor loses one million dollars. To recover the one million dollars the contractor must complete another eleven million dollar project with a 10% profit margin. After completing the second eleven million dollar project, assuming the contractor has made the tendered profit of one million dollars, the contractor has covered their loss from the first project. They've completed twenty two million dollars of work and made no money, yet they still have to pay their overhead costs.

Indeed, it is difficult to recover from losses.

Now imagine the same company has office overheads of a million dollars. (These are the costs to run their office such as the office rental, telephones and salaries). The contractor has to carry out another eleven million dollar project at a 10% profit to cover these costs. At the end of these projects the contractor has completed thirty three million dollars of work and has not made one cent of profit.

Now, imagine another contractor who doesn't lose money on any of their three projects of eleven million dollars each. Further, the contractor operates a small, more modest office, and the office overheads are only six hundred thousand dollars. After completing the three contracts they have made a profit of two million four hundred thousand dollars. (Three projects each made one million dollars profit less the company's overhead costs). This would make many contractors happy! It's been achieved by projects not losing money and by keeping office costs modest.

Of course, if the second contractor was to make an additional 2% profit on each of the projects they would make an additional six hundred thousand dollars (two hundred thousand per project) which would make their total profit now three million dollars.

Consider though, that every project is normally tendered to make a profit, so a project that makes less than the tendered profit has actually lost money. That's right, even though the project has made a profit, as long as it's less than the tendered profit the project has lost money.

Much of what is said in this chapter may seem basic, but it's amazing how many basics are ignored or overlooked. Many items could sound petty, but to put it in perspective, let's consider a company that is undertaking work at a 10% profit margin, so to earn a profit of one thousand dollars the company has to complete ten thousand dollars of work. Therefore by wasting a thousand dollars it's equivalent to having to do an extra ten thousand dollars of work. Which is easier – to do ten thousand dollars of work to cover a loss of one thousand dollars, or to actually not waste the one thousand dollars in the first place?

If you can save a thousand dollars here, and a thousand dollars there, then fairly quickly you've saved ten thousand dollars, which is additional profit that in the normal course of events would have required the successful completion of a one hundred thousand dollar project.

Case study:

I recently had a small renovation done at my house. The contractor doing the ceilings and partition boards employed a subcontractor for the installation. The subcontractor installing the boards arrived at 7am, but the boards only arrived at 9am. When the boards arrived they were too long to be lifted by hand onto the second floor. A fifteen minute rain squall came through and the boards couldn't be lifted off in the rain. The delivery vehicle left without offloading the boards and the subcontractor went home unhappy because they weren't able to do any work. Two people wasted their whole day. The delivery vehicle had to return the following day. Because the boards were too long they had to be cut on the truck before they could be moved upstairs. To compound the problem there were too many boards ordered resulting in 20% of them not being used. In addition the wastage in cutting the boards probably resulted in a further 15% waste.

The end result was that the final labour cost was nearly double, and the material cost was about 35% more than it should have been.

If the correct quantity of material had arrived, when it should have, in an appropriate size, these additional costs could have been avoided.

Yet what I have described is fairly typical for many projects, and it's sometimes surprising contractors make a profit at all.

Work smarter

I always maintain there are many different ways of constructing something, or carrying out a task. Some of these are obviously wrong. However, there are usually a number of different options which are correct. But some are more correct than others. By this I mean that all the correct options will end up with the required end product, it's just that some choices will result in a better organized and safer project, with a shorter schedule, requiring fewer people and less equipment, and will overall be more efficient and cost effective than the other options.

It's therefore important not to pick the first method of construction that comes to mind, but rather to consider various options, weigh up their risks and benefits, and then select the best.

Plan

An important aspect of reducing unnecessary expenditure is to ensure the work is planned ahead of time so that:

- it's not delayed by lack of access
- all materials are available on site when they're required
- the appropriate equipment is available
- the preceding necessary work has been completed
- sufficient and competent personnel are available
- the appropriate paperwork has been completed and submitted (such as method statements, test results and job hazard assessments)
- the appropriate tests and inspections are completed

There are often additional costs due to poor planning because:

- personnel and equipment are standing idle waiting for access, materials or equipment
- management use their valuable time to make emergency arrangements to procure materials, organise access and rearrange work sequences, instead of managing the project
- materials which aren't on site have to be expedited at additional expense
- there are delays to the project schedule
- there's a knock on delay to follow-on trades and subcontractors

Schedule

A properly constructed schedule can save money. This can be achieved by:

- moving the critical path through different activities to find the optimum construction sequence with the shortest overall duration, thus saving on project overheads
- ensuring the different activities are scheduled in such a way they minimally impact the access required for other tasks
- arranging activities so the utilisation of resources is smoothed out and is relatively even, so that resources aren't idle or have to be demobilised and then remobilised again
- sequencing activities so that specialist equipment has continuous use, doesn't stand idle or have to be brought back onto site at a later date (for example try and arrange that all heavy lifts on site are done by a large crane in one visit, or that road surfacing equipment only has to be brought to site once to complete the roads)
- ensuring that the client's milestones are met so that the contractor doesn't incur penalties
- scheduling tasks in the correct sequence so they don't have to be redone later
- taking into account the available resources within the company, as well as those required from subcontractors, so that the tasks can be carried out in the time allowed on the schedule

- allowing for design, design approval and manufacturing times to avoid delays
- allowing for any impacts due to adverse weather, and where possible schedule tasks that may be affected by poor weather to happen in a more favourable season
- allowing sufficient time for the client to approve the contractor's management plans and method statements
- taking into account the time required for the issuing of permits

Access

Lack of access, or poorly planned access, adds to the cost because:

- it delays work and impacts the schedule
- it may result in subcontractors being unable to work resulting in them charging standing time and possibly even moving off site
- personnel may have to stop work to allow other work to proceed, resulting in them being idle, or having to relocate to another work area
- poor access may slow processes down, for example delivery trucks have to use longer or slower routes, or workers have to walk lengthier routes to reach the work area which impacts their productivity
- access may be dangerous, reducing productivity and endangering personnel
- materials may have to be double handled
- larger lifting equipment may be required to access the areas

It's important to plan access at the start of the project by:

- scheduling activities so they don't interfere with each other, taking into account:
    - access to work areas, in particular the requirement for scaffolding
    - the impact of lifting equipment and lifting operations on other activities
    - access for delivery vehicles
    - the installation of large and heavy items
- planning access routes so they:
    - are safe
    - require minimal maintenance
    - take into account future activities and structures
    - are the shortest, most efficient ones possible
- planning storage and stacking areas so that:
    - they don't block or restrict access
    - they are close to the work areas
    - materials can be easily delivered and removed as required

Safety

Safety must be appropriate and relevant to the project.

Poor safety could result in injury, disablement and death as well as additional costs due to:

- it resulting in an accident which may result in:
    - management requiring time to investigate
    - the worker being absent for a period of time while being paid
    - a key worker, such as a crane or excavator operator, or a supervisor, being injured, resulting in other workers being unable to work affectively, or even a section of work standing
    - a key piece of equipment being damaged
    - completed work being damaged, resulting in repair costs and delays to the project
    - an increase in insurance premiums
    - poor morale which impacts productivity
- the project being closed down by the client, or a government body, for safety violations
- fines being imposed for safety violations
- unsafe work conditions which affect productivity

Inappropriate implementation of safety measures could result in:

- potential safety issues not being addressed
- additional expenses being incurred due to unnecessary safety procedures being implemented

Poor safety performance will reflect poorly on the contractor and clients may not award projects to a contractor whose safety is poor.

In many small companies the owner performs tasks on site and there's always the risk of injury. When this happens it could be catastrophic – not only will the owner lose income while they are injured but the project may not be able to continue in the owner's absence. In fact the whole company may come to a standstill since the owner literally holds the keys to everything. Staff, suppliers and subcontractors might not be paid as there's no one to authorise the payments. Projects may stop, and the company will quickly lose its hard fought reputation. Indeed, there is more than one business that has failed due to the owner becoming seriously ill or injured.

Quotes & tenders

Often Project Managers place orders with suppliers that are convenient, or who they know. Sometimes these suppliers or subcontractors aren't briefed properly on the task, requirements and restrictions, resulting later in variations to their quoted price.

To obtain the best prices ensure:

- that at least three quotes are obtained
- that the supplier:
    - has all the tender drawings
    - has the correct specifications
    - is aware of the contractor's terms and conditions

    - understands the project's quality systems and requirements
    - is aware of the project conditions which may impact them, such as specific labour agreements and access requirements
    - understands the safety and environmental requirements if they are doing work on site
    - is aware of what the contractor will provide and what they must provide, so costs aren't duplicated
    - understands the delivery or completion dates
    - has the delivery address
    - knows what guarantees and warranties are required
    - is advised of particular concerns or requirements
    - (if shop drawings are required), understands the requirements for submitting them and the time required for their approval
- quotes are in writing

It's important the contractor doesn't just pass on the client's drawings, contract conditions and specifications without reviewing them to ensure that there are no inconsistencies, and that they are all appropriate to the subcontractor.

Adjudicate quotes and tenders

Often orders are awarded to the cheapest subcontractor, or supplier, who ends up in fact not being the cheapest when the contractor incurs additional expenses to manage them, or because the subcontractor later submits variations for what they perceive are additional works but were items allowed for by other subcontractors in their tenders.

When adjudicating quotes and tenders ensure that the supplier or subcontractor has:

- allowed for the correct product
- evaluated all the applicable drawings
- priced all the items
- met the specifications
- conformed with the schedule
- complied with the warranty periods
- met the required quality standards and documentation
- complied with the site specific conditions
- included all taxes and duties
- allowed for transport
- included their temporary facilities, equipment and services
- adequate insurance in place
- agreed to the payment terms and conditions
- included for preparing of designs and drawings
- allowed for site measurements or providing templates
- included for receiving and handling of materials supplied by the contractor or the client
- not incorporated any unsuitable or unacceptable conditions
- sufficient resources available to do the work

In addition:

- check what additional costs will be incurred for items excluded from the subcontractor's price, or for services, facilities and equipment which the contractor must supply
- take into account additional travel or supervision costs which may be incurred to inspect the manufacturing process
- compare the quotes with the allowances in your tender

The quotes must be adjudicated fairly, allowing for all additional costs.

Negotiate with suppliers and subcontractors

It's beneficial to develop relationships with suppliers by using the same ones on a regular basis, and visiting them to explain who you are, what the company does, your current projects, and how they can assist and benefit from the relationship.

Most suppliers have various trade discounts which are given to companies that regularly purchase from them. These discounts often depend on how much business the company provides. Always ask for a discount.

A large order may result in a cheaper price compared to several smaller ones, so it's always a good idea, where possible, to place the full order at the start of the project, and then schedule the deliveries at intervals over the project duration.

Often suppliers are willing to offer early payment discounts, but then it's important they are paid within the specified period to take advantage of the discount, because paying even a day late may mean the discount is forfeited.

Developing relationships with suppliers and ensuring they are paid on time, will usually result in more efficient service, favourable payment terms (which helps cash flow) and possibly even discounts.

Nevertheless, it's important that contractors continually check that their regular suppliers are providing the best service and cheapest prices, so it's essential to also get quotes from other suppliers.

Orders

Poorly written or incomplete orders can result in the wrong material being supplied, or the project paying more for an item than was agreed.

Orders must:

- be clear and unambiguous
- have the project name
- have the date of the order
- be checked to ensure it's as per the agreed quote
- include an order number
- have the supplier's name and contact details
- include a complete description of the product
- reference any standards, specifications and drawings with which the product must comply
- have any specific manufacturing instructions and details
- specify the delivery date

- include the arrangements to transport the item (if the supplier is providing the transport include the full delivery address, instructions and a map)
- in the case of large items, include the arrangements for who is responsible for unloading and stacking
- have the product price, specifying the unit of measurement and what's included in the price
- specify the terms of payment and trade discounts
- include the address where invoices must be sent
- specify the warranties and spare parts required
- include the name and contact details of the person issuing the order
- be signed by an authorised person (this may depend on the value of the order, since personnel are often allowed to sign orders only up to a particular value, and more senior management are required to sign orders with a greater value)
- be acknowledged by the supplier so there is a record that the supplier has received and accepted the order

Labour only subcontractors

Labour only subcontract orders must be clear and unambiguous so no additional and unexpected costs are incurred. These contracts should specify:

- what's included in the wage costs such as:
    - allowances
    - leave pay
    - paid public holidays
    - bonuses
- overtime rates and when they apply
- who from the contractor will agree the hours worked, and that this is done daily
- who is responsible to supply personal protective equipment
- who covers the costs for mobilisation, inductions and medicals
- who supplies and pays the costs for transport and accommodation
- who pays for replacing unsuitable workers
- who is responsible for implementing discipline
- the insurances required
- the wage agreements the subcontractor must use
- special project rules and working hours

Subcontractor orders

Additional costs are often incurred due to poorly worded subcontract orders which fail to spell out the contractor's and subcontractor's responsibilities, leading to claims, disputes and unhappiness. Subcontractor orders must be clear, unambiguous, without contradictory conditions and clauses and should include:

- the scope of works
- reference to drawings where necessary

- reference to the particular specifications
- the quality procedures, documentation and testing
- the safety requirements
- particular and special conditions pertaining to the project, including specific labour conditions and agreements
- the schedule, including highlighting key dates and any discontinuities in the subcontractor's work
- commissioning requirements
- spare parts, guarantees and warranties required
- payment conditions and procedures for submitting valuations
- mobilisation procedures and requirements
- documentation required before work starts
- warranties, sureties, bonds and guarantees required
- samples required
- shop drawings required as well as their submission guidelines and the approval time and process
- documentation required to close the contract out
- the contractor's right to ask for the removal of unsuitable subcontractor's staff or equipment from the project
- penalties or liquidated damages applicable for non-performance
- the right to vary the subcontract works, including the reduction in scope
- termination clauses
- the obligations of each party
- the contract price (include a breakdown if necessary or reference pertinent rates)
- procedures for lodging variations

The document should be signed by authorised representatives of the contractor and the subcontractor.

Manage subcontractors

Once a subcontractor is appointed they must be managed. Failure to do so can be costly to both the contractor and subcontractor.

Poorly managed subcontractors can result in the following:

- Poor quality workmanship which has to be redone, delays the schedule, and costs money to repair.
- Contractual disputes which cost time and money to resolve.
- If the subcontractor is overpaid the contractor may not recover the money.
- If the subcontractor leaves the project without completing all of their work, another subcontractor may have to be appointed to complete the work at additional cost.
- If the subcontractor falls behind schedule it will impact on the project schedule and other activities.
- The subcontractor may injure someone.

- If the subcontractor uses materials of an inferior quality, or which don't meet specifications, they will have to be replaced, causing delays and adding costs.
- Miscommunication could result in the subcontractor doing work incorrectly, so instructions should be clearly given in writing.
- Failure to give access to the subcontractor on time may result in them charging standing time.
- Allowing the subcontractor to perform work prior to agreeing the price may result in the work costing more than if the contractor had appointed another subcontractor to do the work, or had the opportunity to negotiate with the subcontractor.
- Failure to ensure the subcontractor cleans their work areas results in the contractor incurring this cost, and uncleared rubbish could cause a safety hazard.
- Releasing final payments before the subcontractor has completed all punch lists, and delivered all documentation and guarantees, may result in these being delayed, or not being received at all.

In addition, poor work by subcontractors will damage the contractor's reputation since the client views all work on the project as if it was performed by the contractor, even if it's done by a subcontractor.

Materials

The poorly managed supply of materials can result in wasted and additional costs because:

- they don't meet the specifications or quality requirements (in particular, care should be taken that imported goods meet the local building codes)
- permission hasn't been sought or granted from the client to use a particular product resulting in the material being rejected
- materials of the incorrect specifications are used
- they aren't adequately insured against damage or theft
- they're too large and cannot be installed
- they're too heavy to be handled and installed with the available lifting equipment, resulting in delays and additional costs to mobilise alternative equipment
- the material is difficult to handle, has inadequate lifting points, or is packaged incorrectly
- the item is difficult to fix in position because inadequate steps have been taken to ensure it can be secured and kept in position during construction, resulting in delays and additional costs
- materials are not adequately protected from impact or weather damage during transport or installation, resulting in damaged goods arriving on site which may result in delays
- the incorrect quantity of material is ordered resulting in a surplus or a shortage of material

Coordinate and plan deliveries

Projects often incur unnecessary costs because deliveries aren't planned or coordinated.

These costs are as a result of:

- trucks standing waiting to be offloaded because:
    - suitable offloading equipment is unavailable
    - the offloading area isn't ready
    - there isn't suitable access to the offloading area
    - the documentation, including risk assessments or lifting studies aren't available
    - personnel required to offload are not available
- materials being offloaded in the incorrect place requiring double handling to move them to the correct area
- trucks are turned back empty because:
    - they are the wrong size or type of truck
    - the item isn't ready
- the delivery trucks are too big, or awkward, to access the site
- trucks going to the incorrect delivery or collection address
- the materials not arriving in the correct order, with the materials required first arriving last
- materials arriving late
- materials arriving too early resulting in them being double handled

Reduce waste

At the end of every project I've been involved with there have been surplus materials left over. These are often as a result of:

- the incorrect quantity of material being ordered
- duplicate orders being placed
- material of the wrong specification or size being ordered
- the material supplied was of an inferior quality and couldn't be used

Of course sometimes the surplus material is a result of the client changing drawings after the material has been ordered, in which case the client should be charged for these, and they should be handed over to them.

Other causes of wastage are as a result of:

- Material breakages due to it being damaged during transport, offloading, handling, or using the incorrect installation procedures. Sometimes some of the damage can be avoided by changing the way the material is packaged, handled or transported. For instance if material is palletised by the supplier – which could be at an additional cost – not only can the material be offloaded and handled more easily, thus reducing costs, but there will also be fewer breakages.
- Contamination of materials which is a particular problem with concrete aggregates, or road building materials, which become contaminated

when they're mixed with other materials, or with the ground they're dumped on. Sometimes trucks aren't cleaned properly between products, resulting in the next product being contaminated. Occasionally goods aren't handled correctly at the supplier and become contaminated there.

- Products being applied incorrectly because:
    - they aren't prepared or mixed correctly which results in the work having to be redone
    - the product may be applied too thickly – for instance concrete slabs may be formed and poured too thickly, paint and epoxy coatings applied too thickly, or joints formed too wide requiring additional sealant material
- More product is mixed than can be used, resulting in the unused product having to be discarded.
- Incorrect storage, which may result in materials being damaged by water, dust or heat.
- Keeping materials beyond their shelf life which results in the material having to be discarded.
- Poor housekeeping on a project may result in materials being mislaid or damaged by being walked or driven over.
- Materials are supplied in standard lengths or sizes which usually have to be trimmed to fit where they're required. Often the offcuts can't be used elsewhere. If thought isn't applied to what the best suitable size is then the quantity of these offcuts can be large.

The additional costs to the project are:

- the cost of the additional materials and their transport to site
- offloading, handling and storing them
- the cost to dispose of the additional, broken or contaminated material which includes:
    - handling and loading
    - transport
    - dump fees

Alternative materials

Sometimes there can be savings by using alternative materials because:

- they cost less
- they are easier to install because:
    - they are lighter
    - they require fewer fixings
    - they are easier to handle
    - the size is more convenient
- a different size reduces wastage due to:
    - there being fewer cuts
    - there being fewer and shorter offcuts
    - there being fewer laps

    - them not being as easily damaged during transport and handling
- they result in cheaper transport costs
- there may be fewer breakages during transport, handling and installation
- they may be more durable which later saves on maintenance costs
- they could shorten the installation time, thus shortening the schedule

In addition, there may be other benefits of using alternate materials such as they may:

- be safer to handle and install
- provide a better finish

It's usually necessary to seek the client's approval to use a different material, but if the client incurs additional design costs, they will be reluctant to grant approval. However, it's often in the project's interests, in which case the contractor must demonstrate this, or alternatively offer the client a saving to use these materials.

Reconciliation of materials

Comparing the quantity of material delivered with the quantity invoiced to the client may show a difference, which could be due to:

- the client being invoiced for less than they should be
- materials being stolen on the site, or en-route to site
- materials being wasted
- materials being applied too thickly
- suppliers invoicing for more than they supplied
- materials being mixed in the incorrect proportions

It's important to reconcile materials regularly so discrepancies are detected early, which allows action to be taken to prevent further losses, and possibly even to recover the losses incurred.

Price increases

Price increases are inevitable on most projects with a long duration. It's possible to reduce the impact of these increases by:

- being aware of when the increases will happen:
    - ask suppliers to advise you in advance of when an increase will come into effect
    - many industries adjust their prices at the same time every year
    - being aware of any external factors which may influence the cost of an item, such as volatile foreign exchange rates and increases in labour, raw materials or fuel which could impact the cost of the item
- before an increase takes affect it may be possible to purchase the outstanding material and:
    - store it on site
    - ask the supplier to store it
    - store the material at another location

However, these solutions may mean that material is damaged by the weather, has to be double handled on site, or have additional storage costs, which makes these solutions more costly and unviable.

- when the order is placed it's sometimes possible to request the supplier to fix the price, although this may mean paying a premium

Alternative transport

Transport can be a major component of the cost of items so it's often worth looking at alternatives. There's sometimes an option of using transport collecting material from another company on the project, or close to the site, that's returning empty.

However when arranging transport check:

- the type of vehicle, as some trucks may not be suitable (for example the item may be too wide or long for the truck, or if the truck doesn't have sides it could be unsuitable for the load)
- the capacity of the vehicle (you may think you are paying less for a load when in fact its capacity is less)
- the items are insured
- how easy it will be to offload the vehicle (some covered trucks might have to be offloaded by hand, adding to the costs)
- the delivery times (some trucks could only be scheduled to arrive late, even after hours, or have a longer travel time)
- the roadworthiness of the vehicles (un-roadworthy vehicles might not be allowed onto site and in addition they are at risk of breaking down or being involved in an accident)
- any additional costs such as standing time, or restrictions such as the time allowed to load and offload the vehicle
- that the loads go directly to and from the site, not via a staging area, where they may be reloaded onto other vehicles, with the possibility of damage or theft

Labour productivity

Labour is often a major component of the costs on a project. It sometimes accounts for more than 50% of the costs meaning even a 10% improvement in efficiency can result in an additional 5% profit. Of course the converse is true, and if labour is 10% less efficient than expected then the profit is reduced by 5%.

But it's usually more than just the direct costs of the workers.

Low productivity means more workers are required, which adds additional costs for accommodation, transport, mobilisation and supervision. Poor productivity also impacts the schedule which can result in the client imposing penalties for late completion as well as the contractor incurring additional overhead costs.

Poor labour productivity is sometimes obvious when there are people standing idle on site. However, often the poor productivity isn't picked up before there are delays to the schedule, or the cost reports show labour losses. Usually by then it's too late to rectify the problem.

It's imperative to analyse why there's poor productivity. There's some truth in the saying 'a busy worker is a happy one'. Workers who are idle tend to chat to colleagues, even influencing and interrupting others who are working, and start to see and create problems where there weren't problems before.

Poor productivity could be a result of:

- having too many people on the project
- the area being too congested or cluttered
- poor supervision
- waiting for access to work areas
- waiting for equipment or materials
- having the incorrect equipment or materials
- equipment breakdowns
- insufficient equipment to move and lift materials
- materials difficult to handle and work with
- insufficient resources of one trade and too many of another resulting in one trade waiting for the other to complete their work
- workers aren't sufficiently skilled
- workers may be unhappy because of poor working conditions or clashes with other team members, their supervisor or management, which often results in them performing tasks slowly
- poor discipline
- lengthy meetings with workers or with supervisory staff
- the project isn't planned and coordinated and the subcontractors' and client's activities impact on the work
- poor safety and housekeeping resulting in:
    - lost time due to incidents
    - the project being shut down due to poor safety
    - accidents leading to poor morale
    - key people being injured and unable to work
    - tasks being done more slowly
- fatigue (it's important to ensure workers don't work excessive and long shifts or work on rest days and project breaks)
- a high turnover of personnel which is disruptive, leading to poor productivity because new personnel have to learn the project rules, systems, procedures, and tasks, and work with new team members
- workers are frequently moved between the tasks in the course of the day, resulting in:
    - lost time when they pack up tools and move, as well as going through safety briefings and explanations of the new task
    - reduced productivity since they're unable to gain familiarity with the task or other team members (there's a learning curve to most tasks and as workers become used to performing a task they usually become more proficient and more productive)

- management is indecisive and fail to make timely decisions
- access routes and roads are poorly planned, becoming blocked or restricted
- storage and stacking areas aren't planned and become unsafe or inaccessible
- long distances between work areas, the toilets, offices and stores
- poor performance of subcontractors
- work areas have inadequate lighting making it:
    - unsafe
    - difficult to ensure the correct quality
    - leaving dark areas where workers cannot be observed
- personnel don't work their full hours, they may take extended lunch and tea breaks and arrive at the work site a few minutes late and leave early (often this can amount to thirty minutes for each person, every day, so it's important Project Managers set the right example with their own time-keeping, and enforce good time-keeping on the project from the beginning)
- adverse weather such as rain, wind, high temperatures and low temperatures (steps can be taken to limit the disruption such as; moving workers to other areas, providing adequate protective clothing, allowing sufficient rest breaks, reducing work hours, planning projects to ensure much of the work is done before the onset of poor weather, making buildings weather-tight as soon as possible and lifting items at times of the day when it's less windy)

Sometimes poor productivity is a result of the client. The reasons must be ascertained so measures can be implemented to rectify the situation, or so that variations for the additional costs caused by the client can be submitted. Reasons may be because of:

- late information
- drawing changes resulting in rework
- lack of access or late access
- the client's or their subcontractors' activities impacting the work

It is important that Supervisors and Project Managers are made accountable for their worker's productivity. Labour productivity should be monitored on each project on all sections. Feedback must be provided to Supervisors and workers, and suggestions made as to how the productivity could be improved.

Pay the correct wages

Project Managers must read and understand the industrial relations agreements governing the project. These may include the prevailing labour legislation and project, union and company agreements. Employees' wages and conditions must be in accordance with the relevant agreements which often differ between projects.

Many projects incur additional costs because they don't manage the hours and pay of personnel correctly which could be because:

- the hours personnel work aren't recorded correctly
- people aren't paid the correct overtime rate or the rate is applied incorrectly
- people aren't booked absent when they're not at work
- leave forms aren't submitted (including annual and sick leave)
- people are paid the incorrect rates
- deductions for tax aren't applied correctly and in accordance with the latest legislation
- the incorrect deductions are made from their pay
- people are paid allowances they aren't entitled to
- rates of pay or allowances aren't adjusted when personnel are transferred to another project
- people deliberately falsify their attendance
- errors are made when completing time sheets
- people leave work before their shift is complete

Industrial relations

Management must apply industrial relations policies in a fair and consistent manner in accordance with the project labour agreements and labour legislation (which changes from time to time). Failure to do so can impact the costs of doing business because:

- personnel are overpaid which:
    - adds to the costs
    - causes people who weren't overpaid to be unhappy, affecting their productivity
- people are underpaid which results in:
    - them being unhappy leading to poor productivity
    - management spending time resolving the problem
    - some instances, legal disputes which cost time and money to resolve
- rules aren't uniformly or correctly applied leading to poor discipline which:
    - can affect the project safety
    - takes management time to resolve when there are issues
    - impacts the company's reputation
    - affects productivity
- disciplinary processes aren't carried out correctly, leading to people being unfairly dismissed resulting in:
    - the person being reemployed and paid wages for the time they weren't working
    - legal costs
    - wasted management time
    - poor discipline amongst other employees when they see the dismissed person reinstated

Companies should:

- have suitable industrial policies and procedures in place which comply with legislation
- educate management and supervisory staff on the application of these procedures
- ensure that company policies and procedures are updated when legislation is changed, and that projects are aware of these changes
- make sure all staff apply these policies and procedures consistently and fairly

Control overtime

Often projects work excessively long shifts, or work weekends and public holidays. These longer hours are normally paid at overtime rates which could be anywhere from 50% to double the normal rate. Obviously the workers aren't double, or even 50%, more productive during this time. In fact the opposite is normally the case and as fatigue sets in the productivity declines. In addition, some of the supervision and management may be absent during this time which can result in workers being poorly managed and supervised.

Sometimes when overtime isn't controlled, personnel will work on a weekend when they earn the overtime or penalty rates, and then take a day off in the week which would have been paid at normal rates.

Workers may be poorly controlled on weekends and they arrive on the project intoxicated, or they don't work the hours they claim they have worked.

Another problem when work is carried out afterhours is that key personnel may be absent; for instance, if the crane operator decides not to work then other workers may be unable to do their work.

It's therefore important when overtime is necessary that:

- only work that has to be done after hours is performed
- only those that are required for these tasks are allowed to work
- the hours are carefully logged and monitored
- there is adequate supervision present
- the work areas and equipment are accessible
- key personnel are present
- the hours of individuals are monitored to ensure they don't exceed the legislated hours, or the prescribed hours for the project
- workers are monitored to ensure they don't become fatigued
- workers are paid correctly and in accordance with the prescribed overtime rates
- all arrangements are in place for transport and access to the site
- the client is aware work will be done after-hours
- the project isn't breaking any codes or regulations regarding noise restrictions or similar

Damage to equipment and property

Many projects incur additional and unnecessary costs when equipment or property is damaged.

Damage may be caused when equipment:

- is operated by unauthorised, unlicensed or poorly trained operators
- is operated by operators under the influence of alcohol or drugs
- is used in the incorrect application
- is not maintained correctly
- is overloaded
- has the incorrect parts fitted, or is filled with the incorrect lubricants or fuels
- isn't checked for damaged or worn parts
- doesn't have the oil and lubricant levels checked regularly
- is used in dangerous or unsuitable work areas where something may fall on it, or where it could fall over or become bogged down
- is parked with its keys in the ignition which allows for unauthorised use or theft
- is parked in unsafe areas at the end of the shift where it could become flooded or damaged
- isn't working with a spotter in restricted places
- operators don't check the machine after returning from a break to ensure it's safe to operate, or don't assess what circumstances have changed in the interim which may affect the safe operation of the equipment
- operators don't understand their tasks
- is working in proximity to people and other machines and the operator isn't aware of them
- operators are careless

Furthermore theft and vandalism also results in additional costs, so equipment should be parked and stored in a secure area.

I've also experienced operators deliberately damaging their equipment because they are unhappy with the company, Supervisor or Management or because they are trying to slow the project down.

Tasks must be planned and coordinated to reduce the interaction between different items of equipment which could come into contact with one another.

Put items off hire

Often projects forget to put items off hire when they're no longer required. Alternatively suppliers are verbally notified the item is off hire and there's no written record of this which can result in the item remaining on hire.

Small tools are frequently hired and then placed in the project store, or Supervisor's office when they are finished the task, where they remain for weeks or months, and in some cases even remain in the office when it's transferred to the next project. Over a period of time the hire costs mount and become significant.

Externally hired equipment

Projects often incur additional unnecessary hire charges because:

- breakdowns aren't reported in writing to the supplier
- the breakdown hours aren't recorded on the time sheets
- the time sheets aren't signed off and agreed daily
- hours are booked which weren't worked, such as lunch and tea breaks
- the item didn't work the minimum agreed hours
- the item wasn't checked for damages or worn items, like cutting edges and tyres, when it arrived on the project and the contractor is later charged for these repairs when the item is returned to the supplier
- the contractor undertakes repairs and maintenance which the supplier was responsible for
- additional insurance is added to the hire cost despite the item already being covered by the contractor's own insurance
- the contractor hasn't read and understood the terms of the hire agreement
- the level in the fuel tanks wasn't checked and recorded when the item arrived on the project, yet, the supplier expects the same item to be returned to them with the fuel tanks full (on large equipment these tanks could hold several hundred litres of fuel)

Negotiate reduced rates for inclement weather, low usage, and site closures

When a large quantity of items is on hire, or when there are expensive hire items, it's prudent to agree with the supplier that hire isn't charged when the project isn't working, like over long-weekends or builders' breaks. Alternatively it may be worth putting the items off hire during these breaks and rehiring them when the project restarts.

When placing the order negotiate with the supplier for reduced rates, or even no charge, for periods of adverse weather when the item cannot be used.

Similarly, agree the minimum daily, weekly, or monthly hire hours that the equipment supplier requires and ensure that the project can provide these hours.

Equipment productivity

On some projects, such as earthmoving projects, the hire costs of plant and equipment can be significant and improving its productivity can improve the project's profit.

Poor equipment productivity:

- can adversely impact the schedule
- could result in additional equipment having to be brought onto the project
- causes the available equipment to have to work longer hours
- impacts on labour productivity by:
    - resulting in operators working longer hours
    - requiring additional operators
    - holding up and delaying other activities and workers

Reasons for poor productivity of plant and equipment are:

- the area is too congested or cluttered
- poor or inexperienced supervision
- waiting for access
- waiting for equipment or materials
- equipment breakdowns (for example if an excavator breaks down the trucks it was loading often have to stand)
- poor operator discipline
- lengthy project meetings with operators or supervisory staff
- the project isn't well planned and coordinated
- poor safety
- operator fatigue resulting in accidents or poor productivity
- poor or inexperienced operators (they may operate the machine slowly or not at its full production)
- a high turnover of operators
- having the incorrect machine, or one which is underpowered for the task
- using machines for inappropriate tasks (I often see loaders on site being used to cart minor materials around which could easily be transported in a wheelbarrow or utility vehicle)
- machines are frequently moved between tasks, resulting in lost time due to the travel time and re-orientation of the operator
- management is indecisive and doesn't make timely decisions
- access routes and roads are poorly planned, becoming blocked or restricted
- work areas have inadequate lighting
- operators don't work their full hours, often taking extended lunch and tea breaks, arriving at the work site a few minutes late and leaving early
- adverse weather such as rain which causes machines to get stuck and roads to become slippery
- the item isn't rigged or set up correctly, or isn't using the correct set of tools or attachments (for example cranes parked in the wrong position or with the incorrect rigging will slow down lifting operations, and excavators with the wrong type of bucket may damage the machine and impact the time to do the task)
- mismatch of pieces of equipment that are dependent on one another (for instance, the correct size and number of trucks must be matched to the size of the loading excavator as well as to the task and site conditions)
- poor scheduling of refuelling and routine maintenance

Theft

Theft can be a major problem in some areas and it's important to have sufficient security in place. Theft results in:

- the direct costs of replacing the items

- costs to transport and handle the replacement items
- the item being unavailable (for example, personnel who have their tools stolen are usually unable to work, and the theft of an excavator's battery could result in the excavator standing for hours, even days, waiting for a replacement)
- delays to the schedule
- specialist tradespeople may have to return to the project, at additional cost, to install the replacement item

Prevent problems from occurring

A large portion of additional costs on a project could be avoided if the contractor's management team anticipated problems, and implemented steps to prevent them from arising. Some of these problems may be caused by:

- errors with the drawings, or the information supplied by the client
- drawings or information issued late
- approval of shop drawings or designs by the client taking longer than they should
- materials not arriving on time
- failing to order materials
- poor performance from subcontractors
- the incorrect material or poor quality material being supplied
- access restrictions
- poor quality work
- insufficient resources
- industrial relations problems
- safety concerns
- material handling difficulties

Many of the issues can be prevented if the contractor plans the project correctly, choosing a suitable construction methodology, planning access, preparing an effective schedule and having sufficient and suitably qualified personnel. Adequate preparation for the project as well as for individual tasks will also help prevent problems from arising.

Alternative designs

Consider alternative designs which can:

- save time and shorten the schedule
- be safer to construct
- be constructed using fewer resources
- be constructed using equipment the company owns or that is readily available
- be constructed using fewer skilled personnel, or using personnel with skills that are readily available
- achieve a better quality
- use materials that are readily available or cheaper

For instance by changing elements of concrete structures to precast units it may be possible to build most of them off-site and install them using fewer skilled people. Sometimes a relatively minor change, like changing column sizes to suit standard form-work, can achieve savings with minimal impact on the design.

Some changes may be more significant and could entail the contractor incurring design fee costs and accepting some of the design risk. It's necessary to carefully weigh up the benefits versus these additional risks and costs.

Coordination of services

On many projects coordinating the installation of services can be important so that:

- services don't clash which often results in:
  - delays while alternate solutions and routes are considered
  - additional costs when the service that has already been installed has to be rerouted to accommodate the other services
- the services that needs to be installed first, for instance the deepest underground services, are installed before the other ones
- when underground services share a common route they are installed together before the trench is closed
- structures take the services into account and allow for all penetrations and other requirements
- in some cases, it may be possible to make use of the same access scaffolding

Install services correctly

Poorly installed services often result in quality problems during construction, or, in some cases, long after project completion.

To prevent problems services should:

- be installed in accordance with the project specifications
- be installed by suitably qualified tradespeople who should be managed by knowledgeable supervisors
- be tested before they are covered up to ensure there are no leaks or damages
- have suitable corrosion protection
- be installed in the correct position and level
- be firmly secured so they cannot work loose
- be clearly recorded on as-built drawings, and where possible on the ground, so they can be located by follow-on trades so that they aren't damaged

Poorly installed services lead to:

- additional costs to repair them
- burst water pipes which can flood buildings causing further damage
- other completed work being damaged when the repairs are carried out
- embarrassment with the client

- the client's operations being disrupted
- dangerous situations, particularly with faulty electrical cables or burst high pressure pipes or gas mains

I've often incurred huge costs redoing finished areas damaged when poorly installed services had to be repaired, or even in some cases, because the service had been accidently omitted.

Damage to existing services

Existing services are often damaged when excavations are being carried out. In fact clients often joke that if they don't know where the underground services are they should call a contractor in, and they'll quickly dig them up. Of course this isn't so funny for the contractor, especially if they damage a fibre optic cable which costs thousands of dollars to repair. In fact contractors definitely don't want to damage existing services because:

- the repair costs can be large
- when a high voltage electrical cable is damaged there's a risk someone will be electrocuted
- work is often held up until the repair is completed, which may take several days
- often a cable is unknowingly damaged, but remains operational, and only later, after the contractor has installed concrete floors and finishes the cable fails, resulting in the completed work having to be broken up to repair the fault
- if an electrical cable feeding the site is damaged it could mean the site is without power for several hours or days, impacting the schedule and productivity
- the client and neighbouring businesses will be unhappy if the contractor disrupts their services (even claiming for losses incurred)

It's important that the contractor:

- obtains all the necessary permits before excavating, drilling, cutting into, or demolishing structures
- ensures they have located all the known services
- clearly marks these services (I cannot tell you how many times we've damaged known services because they weren't marked)
- protects the services where possible
- ensures all newly installed services are clearly marked on the ground and on drawings
- ensures workers are aware of the services and take precautions not to damage them
- clearly highlights the risk of damaging the services in risk assessments and at prestart meetings
- uses personnel that are adequately trained and competent, so they don't accidently operate a machine in a way that damages a service

Damage to services applies to overhead ones as well, and the contractor must take steps to ensure that construction machines don't come close to overhead power lines or structures, as this could result in a fatal accident.

Protection of existing structures and new work

Projects regularly incur additional costs because new work is damaged by the follow-on trades. This damage results in:

- additional repair costs
- personnel being taken off other tasks to carry out the repair which may impact the schedule
- possible delays to the handover of sections (particularly if a replacement item has to be ordered, and it has a long-lead time)

Case study:

On one of my projects someone damaged a roof sheet while working on the completed roof. The hole was three centimetres in extent but the whole sheet had to be replaced. The problem was the roof sheet was about forty metres long and had been rolled on site by a specialist machine which had since left and was available only several months after the project was completed. In addition, because of the way the sheets were laid, several of them had to be removed to replace the damaged one. Carelessness on the part of one worker cost us several thousand dollars to repair the damage.

To reduce the chance of damaging completed work:

- sequence the work so items that can easily be damaged are installed as late as possible, and tasks which require heavy equipment as early as possible
- all workers should be encouraged to take pride in their work and that of others, taking care not to damage the finished product
- protect finished items where possible
- prestart meetings should reiterate the need for care, and highlight methods to be implemented to protect the finished work

Quality

Poor implementation of quality management systems and poor workmanship can result in additional costs because:

- work has to be redone
- it delays the schedule while defective work is replaced which:
    - may result in penalties
    - increases overhead costs
    - impacts on follow-on trades and tasks
- it's expensive to remove waste generated by demolishing defective work
- redoing work takes resources away from other tasks
- redoing tests that have failed costs money and causes delays
- when defective work becomes apparent after the project has been completed the contractor has to return to the site to carry out repairs

(often these disrupt the client's activities and consequently have to be done after hours, making it even more expensive)
- the incorrect tests are performed, or tests are done more frequently than required
- tests have to be redone because they weren't witnessed or documented correctly
- the repairs aren't done correctly resulting in them being redone
- the quality systems and documentation are too complicated and cumbersome, which takes additional personnel to implement, and delays the schedule

To minimise additional costs caused by poor quality, or the failure to implement quality systems correctly:
- a project quality plan must be in place which complies with the project requirements and legislation
- quality systems and documentation should be implemented from the start of the project
- management shouldn't tolerate poor work
- all personnel must understand the acceptable quality standards and take pride in their work
- suitably skilled personnel must be employed
- suppliers and subcontractors must understand the quality requirements and documentation required, and must be able to satisfy them
- ensure that the work and materials comply with the specifications and quality requirements
- testing must be done in accordance with the specifications and quality plan, and must be done by appropriately trained and qualified persons, using the correct equipment, calibrated in accordance with the supplier's recommendations, and witnessed by the appropriate persons
- repairs of defective work must be done correctly
- setting out must be done in accordance with the drawings, and must be checked against existing structures, and if there are doubts about the correctness of the setting out, then questions must be raised
- the most recent drawing revisions must be used and personnel must correctly interpret drawings and raise any concerns or uncertainties they may have about the information
- workers must have access to the appropriate tools and equipment to be able to meet the quality requirements
- work methods and procedures should be implemented to make the construction as simple as possible, and to where possible eliminate the possibility of poor workmanship
- materials must be stored and handled correctly to prevent them being damaged
- materials and equipment must be installed in accordance with the manufacturer's instructions
- adequate lighting must be provided in the work areas

Personnel must understand the consequences of their mistakes. The importance of good quality work must be stressed from the start of the project.

Of course, sometimes it isn't necessarily a quality problem but rather the client has unrealistic expectations about the quality they will get. This can be anticipated and avoided by:

- submitting samples before work starts and obtaining the client's approval for their use
- preparing mock-ups or prototypes for the client to approve
- getting the client to accept work as it proceeds so corrective measures can be taken early if it's not to the required standard
- when the client has specified cheaper finishes or materials, explaining to them that these could compromise the quality of the finished product

Non-conformance reports must be prepared for all defective work to ensure that the defect is remedied correctly and the problem doesn't reoccur. It's also useful to track the cost of defective work as many staff don't appreciate how costly their mistakes are.

Quality documentation

Not carrying out the required tests and inspections, or not keeping records of these inspections, can be expensive.

Case study

Our company constructed a major new airport. Part of the contract included the installation of several kilometres of fuel lines from the fuel storage area to the aircraft parking areas. The line consisted of twelve metre long steel pipes welded together. A few months before the airport was due to be completed the client asked for the quality documents for the fuel line. As part of the quality procedures each of the pipe welds was x-rayed. However the x-ray records couldn't be found.

The fuel line was a critical component and leaks would create a safety hazard, as well as an environmental issue and result in the loss of expensive fuel. The client insisted each of the pipe joints be exposed and re-x-rayed. By this stage the line was closed up, and much of it passed under the completed concrete taxiways and aircraft aprons.

The excavation, re-x-raying and reinstating of the areas cost several hundred thousand dollars. Additional resources were brought in from other projects to ensure the task was timeously completed and didn't delay the opening of the airport.

In all of this, rumours in the press abounded, and the bad publicity continued for months after the project was completed.

The quality documentation must:

- be simple and easy to carry out
- be done as the work proceeds
- comply with the specifications and the client's requirements

- not be done just for the sake of completing paperwork; the person completing it should physically inspect the items to ensure their compliance with the specification
- have a copy retained by the contractor at the end of the contract (I have had cases of the client misplacing their copy resulting in us having to replace it)
- make provision for reporting, investigating and tracking non-conformances and deviations
- include copies of all tests

Ensure repairs are done correctly

I've often seen repairs and rectification work done hurriedly, sometimes even in secret – after-hours. This is obviously unacceptable and results in additional costs.

- Repairs must be done correctly so that the structural integrity of the structure isn't impacted and in such a way that it's aesthetically acceptable. Poorly executed repairs have to be redone, either during the construction period or, sometimes, long after the project has been completed.
- Usually the repair process must be approved and witnessed by the client's representative and they'll often insist the work is redone if they haven't approved the method.

There are often different ways of fixing a problem and the contractor must select one taking into account:

- the aesthetics of the repair (it's often cheaper to do a patch job, which may however look terrible and be a constant reminder to the client of the contractor's mistake)
- the cost of the repair, although cheapest isn't always best
- the durability of the repair (you don't want to have to return to fix the problem again)
- delays the repairs cause the project
- disruptions to the client's activities
- the client's acceptance of the repair method
- the structure's integrity

Sometimes the best solution is to demolish the structure completely and rebuild it correctly.

Modifying existing structures

Modifying and cutting into existing structures is inherently unsafe since:

- existing services may be damaged
- the structure can be weakened, leading to collapse
- partly demolished structures can be unstable while they are made good, or until new structures are completed (severe weather, additional loads or vibrations may lead to their collapse)

- the machines carrying out the cutting or demolishing can cause excessive vibrations, or impose additional loads on the structure resulting in damage or catastrophic failure
- the impact of falling debris onto existing structures can cause excessive loading and damage
- falling debris can cause injury

To prevent accidents care must be taken to ensure:

- all services are located and protected, isolated or rerouted
- an engineer checks the strength of the structure to ensure it will be structurally sound while sections are being demolished and rebuilt
- only areas approved by the designer are cut or demolished
- the remaining structure is adequately braced to allow for construction activities as well as for adverse weather conditions
- the remaining structures and exposed areas are protected to avoid weather damage to the existing facilities
- once all work is complete an engineer checks the structure before the temporary bracing or props are removed
- work is done by suitably qualified personnel who are adequately supervised and aware of what is required and the precautions that should be put in place
- areas are clearly demarcated and barricaded both during demolition and while there are potentially unsafe areas, so that people and equipment aren't endangered

Inspect existing property

Often work involves building new structures within an existing facility. At times access to the site is along current public or private roads. Some projects are adjacent to properties which may be affected by construction activities. In these cases there's always the risk of the existing facilities being damaged by the construction activities. The owners will always expect their property to be in the same condition, or better, after the construction is complete.

To ensure the contractor isn't liable for damage they didn't cause, it's important that a pre-construction survey and inspection is done to record existing defects and damages. This report should be handed to the owners before work begins, so that they have the opportunity to corroborate the defects.

In addition, the contractor should notify the owner if they see someone damaging the facilities, especially if there's a possibility the owner may think the contractor has caused the damage.

At the end of the project the contractor should repair any damage that they caused, and then request the owner to inspect their property. If the owner is happy they should sign an acknowledgement that they're satisfied with the condition of their property and the contractor isn't liable for damages.

Photographs with date imprints are an invaluable form of recording both the pre-construction inspection and the post-construction inspection.

Survey of existing structures

Often contractors work on structures built by others. It's important to verify that these have been constructed in the correct position and to the correct height. Some of these checks may include:

- verifying bolts and other imbedded items are the correct size and are in the correct position
- checking excavations are the correct size, depth and in the right location
- ensuring earthworks terraces and embankments are at the correct level and location
- making sure the area is safe – this particularly applies to deep excavations
- checking that the structures comply with specifications
- noting pre-existing damages to the structures

Failure to detect and report these errors in writing may result in the contractor redoing the work at their own cost. In addition, if earthworks levels are incorrect the contractor could use more materials than they had allowed.

Before work starts the client must be notified in writing of the discrepancies and given the opportunity to undertake the appropriate rectification, or instruct the contractor to make the necessary modifications.

Avoid survey and setting-out errors

Probably some of the most expensive errors I've experienced on projects have been due to setting-out errors. These errors can be caused by:

- misreading or misinterpreting drawings
- using equipment which has been damaged, not checked for accuracy or not calibrated
- using the wrong setting out marks
- using the incorrect information
- not checking the accuracy and position of existing structures
- inexperienced staff
- drawing errors
- arithmetic errors

These errors can result in major costs when structures in the wrong position have to be demolished and rebuilt, or follow on structures have to be modified to conform to the wrongly built structures. Often this also delays the project.

To avoid errors ensure that:

- all setting-out is double checked
- survey instruments are regularly checked and calibrated
- personnel responsible for the setting out are competent
- setting-out information is clearly given, preferably on a drawing, to the person doing the work

Drawing control and management

Many errors occur because supervisors and subcontractors use the incorrect drawings, the wrong revision or superseded specifications.

This results in:

- delays when work has to be redone or the correct material hasn't been ordered
- additional costs to redo the work
- poor morale of personnel because they have to redo work
- clients becoming frustrated

Refer to Chapter 5 for mitigating measures.

Avoid fines

Projects Managers must ensure the project has all the environmental approvals, construction permits and licenses in place. Without these work could be stopped and fines imposed.

There should be strict compliance with all of the permit conditions, local bylaws and safety and environmental legislation. Failure to comply could result in fines as well as work being stopped, which is costly and results in delays.

Often permit conditions and legislation vary between regions, states and countries, so the Project Manager must be familiar with the conditions that pertain to each project.

Fines can be incurred on public roads due to speeding, overloading, incorrect loading, expiry of vehicle registrations and parking in restricted or limited time zones. All staff should be aware of their obligations when operating company vehicles and adhere to the road rules which may vary between regions.

Overloading not only results in a fine but the truck is usually impounded until the load is sufficiently lightened to comply with the legal limits. This means some of the load must be transferred to another truck which results in:

- the delivery being delayed
- the transport company charging standing time while the truck is impounded
- items possibly being damaged or going missing during the reloading process
- the requirement for another truck
- lifting equipment and personnel having to go to where the truck is impounded to assist with transferring the load

Close out projects correctly

Failure to complete projects correctly, including completing all paperwork and documentation, results in increased costs due to:

- the project extending beyond the contractual end date, often resulting in the application of penalties or liquidated damages
- resources remaining on the project longer than budgeted for

- resources not being released to other projects where they're required, which negatively impacts those projects
- retention money and sureties being retained longer by the client
- insurances remaining in place for a longer time
- the contractor incurring additional overheads, such as the costs of facilities, accommodation, vehicles and security remaining on site longer

It's therefore important that the completion of the project is well planned and coordinated so that all items are completed as soon as possible.

Close final accounts

When a project is completed it's essential that the final account and contract sum is agreed with the client as soon as possible. This includes agreeing all variations, as well as deductions, charges and penalties which the client may wish to impose.

Often, as soon as work is complete, the client and contractor transfer staff from the project and it sometimes becomes difficult to contact the people with the knowledge and authority to agree the final account. This could result in the process dragging out over several months, or longer. In addition, when the site offices are demobilised contract documents get muddled and even lost, or they are archived and become hard to retrieve, making it difficult to compile the final accounts.

In the chaos of demobilising some items may be overlooked and not charged to the client.

The delay in finalising the accounts affects the contractor's cash flow because the client cannot make final payment before the value has been agreed.

Punch lists and maintenance

Punch lists should be drawn up and attended to as soon as possible after the work is completed. Failure to do this:

- often delays the completion of the project
- results in personnel remaining behind after the project is completed to attend to the items, resulting in additional costs which include overheads such as accommodation, transport and supervision
- delays the release of retention money and sureties
- extends the contractor's warranty period resulting in them being liable for defects for a longer period
- causes the personnel left behind attending to the items to become frustrated

The client should be advised of items which aren't the contractor's responsibility.

Expense and vehicle claims

Expense claims should be checked to verify that the purchases are legitimate costs, the receipts are attached, and the expenses are correctly cost coded.

Vehicle claims must be checked to ensure that the vehicle has only been used for business purposes and approved private use.

Recovery of deposits

Contractors often pay deposits for accommodation, services, permits, and sometimes, even for hiring equipment. A schedule of these deposits should be kept to enable them to be recovered at the end of the project. Often the people who authorised the payment of the deposits leave the project before the end, so it's important that staff closing the project are aware of what deposits should be recovered and who they should contact to arrange for their payment.

Plan the release of resources

As projects near their end, resources such as equipment, personnel and staff have to be moved off site. This has to be done in a planned, controlled manner so that the project doesn't have unutilised or underutilised resources which add unnecessary costs. But, it also needs to be done in such a way that the project is completed as quickly and efficiently as possible.

Removing externally hired equipment from the project is relatively easy, though the following should be done to avoid additional costs:

- ensure that the item has completed all the tasks it's required to do
- notify the supplier in writing of the date that the item will be off-hired (giving them sufficient notice in accordance with the hire agreement - some items may require a week or more notice)
- arrange transport to remove the item
- ensure that the item is cleaned and has all its parts and attachments ready for collection
- ensure that staff are aware of when the item will be collected
- timeously notify the supplier should plans change

Internal equipment is slightly more complex because it's in the company's interests to ensure that it's usefully employed, either on the project or elsewhere. The project should timeously notify other projects and the internal hire department that the item will become available. Unfortunately most Project Managers think only of their project, and not of the company as a whole, and often place internal items of equipment off hire without prior notification.

Companies should maintain a schedule of their major plant and equipment showing the project the item is currently on, the expected release date, and where it's planned to move to next. This allows other Project Managers to see what's available and assess if they require the item. This schedule also enables the tender department to see what equipment is available which could affect the projects they tender on as well as their methodologies and pricing. For this schedule to be effective it must be regularly updated, and Project Managers need to provide accurate information regarding their requirements and their equipment release dates.

If an internal item of equipment doesn't have another project to go to the Project Manager should ensure, where possible, similar pieces of externally hired equipment are released from their project first.

The company should also maintain a schedule of its staff with their names, occupations, their current project, when they'll be released and which project they are scheduled to move to. This schedule assists the tender department since they are able to see who is available for new projects.

Project Managers need to notify the human resources department and other Project Managers in advance of when they will be releasing personnel, and what their skills and experience are, so they can be transferred to other projects.

If the company doesn't have a position for a person then the project will have to terminate their employment. Since this process can take several weeks, depending on notice periods and union involvement, it's important that it is started as soon as possible so people aren't idle on the project.

When termination is necessary:

- ensure senior management and the human resources department are advised of the process
- if necessary, advise the client or unions
- ensure contract employees are terminated before permanent employees
- ensure the process is:
    - in accordance with legislation, the person's employment contract, and relevant industrial relations agreements
    - a last resort when there aren't other opportunities on the project, or within the company, including the option of redeploying the person in another role
    - fair
- give the appropriate notice to the individuals concerned
- ensure the correct termination documentation is prepared
- make sure the full pay is calculated, including all notice pay, bonuses and leave pay

Wrongful termination may result in extra costs, including additional pay and legal costs, as well as having the potential to cause industrial relations problems on the project.

Minimise weather losses

Rain, wind and storms can cause damage to the project which delays the works and costs money to repair.

Losses can be minimised by taking suitable precautions such as:

- securing the work area at the end of the shift, or when severe weather is expected
- protecting equipment and materials which may be damaged by rain or wind
- protecting open excavations with berms to prevent storm water runoff entering them

- ensuring drainage is planned and maintained to steer storm water away from the work area
- removing equipment from areas which may become flooded

It's important to note that storms can sometimes occur without warning, or rivers rise unexpectedly. We had an instance when we returned after a weekend to find the river had risen significantly, flooding our work areas and submerging some of our equipment.

Manage the use of company assets

Irresponsible use of company assets can result in large costs. These costs include staff using mobile phones for personal calls, company vehicles for private travel or company internet access for personal business. Sometimes projects incur replacement or repair costs because assets like telephones, computers and company vehicles are stolen or damaged in accidents. It's important that these assets aren't placed in situations which expose them to theft or damage.

Often the individual costs can be small, but in companies with a large staff complement the cumulative costs can be high.

Some companies set limits on the kilometres staff are allowed to travel in company vehicles, or limits on mobile phone usage in a month. Usage however can vary between different projects and these restrictions may need to be revised on a regular basis.

One company I worked for circulated a list with the company vehicles, mobile phone, and internet costs, arranged from the highest to the lowest with the person's name, the asset number and the project name, to Project Managers and Directors. It was then easy to see who was using assets excessively, and by checking the previous month's lists it was possible to see who the serial offenders were.

Communication with staff – buy in

Monitoring and reducing costs isn't something that can be left entirely to the Contract Administrators or Project Managers. Rather it's something that all project staff should be aware of so that measures can be implemented to save costs. It's important that Supervisors are made aware when their section is losing money and what steps need to be put in place to rectify the situation.

Summary

By reducing costs it's easily possible to increase the profitability of the company. Ways in which costs can be reduced are:

- working smarter
- planning projects carefully
- implementing the most suitable and efficient schedule to meet the client's requirements
- ensuring that there is access to the work area
- performing the work safely

- obtaining a number of quotes in a process where the supplier is fully conversant with the requirements of the job
- adjudicating these quotes to ensure that they have allowed for and included all items and costs
- giving suppliers and subcontractors clear unambiguous orders and contracts
- managing subcontractors to ensure that they perform their works in accordance with their contract
- ordering the right quantity of the correct materials, and ensuring they are delivered to the project on time and that the project is ready to receive them
- investigating alternative materials which may be cheaper , easier to install, or less expensive to transport
- considering alternative designs which may improve the constructability of the project
- arranging suitable transport for materials and equipment, and coordinating deliveries
- considering alternative forms of transport
- regularly reconciling the materials delivered with those invoiced to the client and investigating any discrepancies
- maximising both the labour and equipment productivities
- ensuring industrial relations policies and discipline are implemented in accordance with legislation, in a fair and uniform manner across the whole company
- paying personnel correctly
- controlling and limiting the amount of overtime worked
- protecting property and equipment from being damaged
- putting equipment off-hire when it's no longer required
- paying the correct hours and rates for hired equipment
- negotiating discounts, better rates and payment terms for equipment and materials
- ensuring that projects comply with legislation and traffic ordinances to avoid paying fines
- preventing theft
- anticipating and preventing problems from occurring
- coordinating the installation of services and ensuring they are installed and tested correctly
- avoiding damage to existing or new services
- protecting existing and new work
- ensuring structures are set out and constructed in the correct location
- working to the latest drawings
- ensuring all work and materials comply with the required quality standards and specifications, and that suitable quality systems are in place

- planning the close out of projects to ensure all structures are completed, and all documentation is submitted in accordance with the client's requirements
- recovering deposits at the end of the project
- planning the demobilisation of resources
- managing the use of company assets to maximise their effective utilisation
- ensuring all staff are aware of how important it is to avoid unnecessary costs

Chapter 7 – Maximising Revenue

The best way to ensure projects are profitable is to ensure that the client pays for all the work that is undertaken, and it's paid for as soon as possible after it's been completed. This seems obvious, and yet I've found that many Project Managers, Project Directors and business owners aren't good at submitting interim valuations on time, getting the final accounts agreed with the client, or submitting variations correctly and in terms of the contract.

Monthly valuations

Valuations must be submitted on or before the agreed due date, addressed to the correct person, be in accordance with the contract and include the required supporting documentation and calculations. The incorrect submission may result in the contractor not being paid on time, sometimes even the following month, impacting the contractor's cash flow.

The Project Manager needs to understand the basis of the valuation which may be:

- measuring the actual work completed in the month
- measuring the progress against the contract schedule, and then claiming the percentage complete
- claiming the value of a milestone when it's achieved

Where possible the projects should maximise the revenue claimed in the monthly valuations by:

- making sure that all work is claimed
- over-claiming where possible
- ensuring milestones are met so that they can be incorporated in the payment
- making the valuation date as late in the month as possible
- ensuring that variations and claims are submitted and approved quickly so that they can be claimed
- claiming for unfixed materials where applicable
- claiming as much of the indirect costs or preliminaries as possible

Remember, it's better for the contractor to have the money sitting in their bank account than in the client's. However, when compiling cost reports, ensure the over-claims in the valuation are excluded from the revenue used in the report.

Variations

Most contracts will change and vary from the works that were originally tendered for. Contractors must ensure that they are paid for all additional work. Variations can be due to the following:

- additional scope
- errors and omissions in the document
- changes to drawings
- delays due to:
    - late access
    - late information or drawings
    - the client making changes to the completed works
    - unforseen weather conditions
    - unforseen project conditions
    - the client's contractors or workers impeding or preventing access
    - the client-provided services being unavailable, being of insufficient quantity, not provided to the point specified, or provided late
    - the client-supplied equipment or materials arriving late, in insufficient quantities, or not to the correct quality
- changes in specifications
- changes in working conditions such as:
    - unexpected ground conditions, for example, rock
    - encountering hazardous materials
    - the discovery of artefacts
    - the unexpected presence of underground water
- changes of commercial or contractual conditions
- the client delaying or changing milestone dates, or requesting the schedule to be accelerated
- drawing errors and drawing coordination problems
- changes of law within the state or country
- the client or their contractors damaging completed work

It's vital to read through and understand the contract and tender documents, as well as the related correspondence. It's important that the contractor regularly compares construction drawings and specifications with those issued at tender stage to ensure that they haven't changed.

Continually ask the questions:

- 'Are we constructing what we tendered for?'
- 'Are the site conditions as expected at tender stage?'
- 'Has the client fulfilled all their obligations in the contract?'

Sometimes the variation is as simple as submitting a revised rate for an item or task because the description of the item priced at tender stage has changed, for example because:

- the height or dimensions have changed
- the quantity has changed

- the specification is different

The client must be notified of variations as soon as the contractor becomes aware of them, and certainly within the time specified in the contract. Failure to do so may mean the contractor loses their right to claim.

The contractor should ensure that the person preparing the variation has the required knowledge and experience to prepare the claim. If there's any doubt as to the validity of the claim, or what should be included, seek advice from experts within the company or from outside providers. The cost of getting proper advice is often far outweighed by the revenue that can be earned by an expertly formulated and drafted claim.

Preparing and submitting variations and claims

Variations and claims should have as the minimum:

- a description of the event
- the cause
- the date of the event
- the event's impact
- steps taken to mitigate the impact
- the cost and time impacts of the event
- the supporting documentation attached, or refer to supporting documentation referencing:
    - contract clause numbers
    - relevant drawing numbers
    - the pertinent schedule item numbers
    - correspondence relating to the event
    - the relevant specifications

  It's essential that this supporting documentation is relevant, supports the claim and isn't contradictory (any contradictions need to be explained).

The claim must be:

- lodged within the time frame specified in the contract
- addressed to the correct person
- delivered to the correct address

As part of formulating the impact of the event all calculations and schedules should be included. The claim schedule should reference the approved contract schedule. Calculations should reference where the facts and figures came from and how they were put together, and they should be checked for arithmetic errors.

It should be noted that claims must be well thought through to ensure all possible costs have been included. It's very difficult, if not impossible, to go back later and submit a revised claim requesting more money. Where possible discuss the claim with the relevant staff to hear what ideas they have, and whether they can suggest other costs or opportunities that may have been overlooked.

Costing variations

Variations should include:

- labour costs including:
    - the base wages
    - overtime
    - non-productive time such as paid breaks, travel time, time to prepare hazard assessments, inductions, and so on
    - allowances
    - statutory levies such as for training
    - leave pay, bonuses, sick leave, pension, and so on
    - personal protective clothing
    - personal small tools
    - accommodation (if applicable)
    - travel (where necessary)
- material costs including:
    - the actual material cost
    - transport of the material to site
    - offloading and handling
    - protection and packaging
    - quality procedures and tests
    - wastage due to breakages and cutting
    - cutting (unless this has already been included in the labour cost)
    - fixings
    - royalties
    - insurances
    - duties and taxes
- equipment including:
    - hire costs
    - unproductive time
    - mobilisation and demobilisation costs
    - insurances
    - fuels and lubricants
    - wearing parts such as cutting edges, drill and moil points
    - maintenance
    - supporting vehicles such as fuel and service vehicles
    - attachments and ancillary items
- demolishing existing structures including:
    - loading and transport of waste materials
    - temporary supports and bracings
    - dump fees
- supervision and management costs including:
    - the basic salary
    - allowances
    - leave pay and bonuses
    - accommodation and transport
    - computers and mobile phones

- off-site staff such as Contract Administrators, and Planners
- project insurances and sureties
- costs of permits
- profits and overheads
- access equipment
- offices and facilities
- additional security
- protection of existing and completed structures
- additional design and drawing costs
- subcontractors' costs including the contractor's mark-up

It's also important to assess the effect the variation has on the schedule which may result in:

- the overall project duration being extended
- the variation impeding or delaying other works
- the variation requiring resources from other tasks which then impacts on their completion dates

Assist subcontractors with claims

Some subcontractors are unsophisticated when it comes to submitting and preparing variations. When these variations are due to changes or additional work requested by the client it's usually in the contractor's interests to assist subcontractors with these claims since:

- the more revenue the subcontractor receives the more likely it is that they'll be profitable, making them happy and more prepared to provide good service and less likely to lodge claims against the contractor
- in most cases the contractor gets additional profit on the subcontractor's claim

Site instructions

Clients often request contractors to undertake additional work, or change work already completed. Many of these requests are verbal. Contractors must ensure that instructions:

- are in writing
- are signed by an authorised representative from the client
- are clear and unambiguous
- are only accepted by the contractor's authorised representative
- allow the contractor to claim for additional costs or time caused by the instruction
- have wording acceptable to the contractor

Sometimes wording on instructions can be ambiguous making it seem that a section of works is being redone because of the contractor's poor quality, or fault. When the contractor submits a variation the client's Contract Administrator may be reluctant to pay the contractor for the additional work since it appears from the instruction the work is due to a failure on the contractor's part.

Often clients use their standard site instruction form which has a printed statement to the effect that there'll be no cost or time variation for the instruction. These clauses should be struck out by the contractor because at this stage they're usually not in a position to understand all the implications of the instruction.

Normally a site instruction should be priced by the contractor before they proceed with the work.

Delays

The full costs of delays are seldom all considered. The delay could result in:

- the contract schedule being extended, meaning that the contractor remains on site longer than allowed for and incurs additional costs for:
  - the site facilities
  - staff salaries and associated costs
  - the extension of the bonds, sureties and insurances
  - equipment
- the inefficient and unproductive use of personnel and equipment which:
  - cannot be used at all
  - are only partly utilised
- the work moving into a season with unfavourable weather conditions which wasn't allowed for in the schedule or tender, causing further delays and inefficiencies
- material prices increasing in the interim
- the activity happening when other contractors are working in the area which then adversely impacts productively
- an activity being undertaken out of sequence which may result in:
  - access being limited when it's completed
  - the work area becoming congested due to other activities happening simultaneously
  - specialist equipment, subcontractors or personnel not having continuity of work resulting in them having to return to site at a later date entailing additional mobilisation costs and in some cases the equipment or subcontractors may not be available when they are required again, which could result in further delays
  - damage to works already completed
- materials which have been ordered having to be stored because the site isn't ready to use them, resulting in storage costs and double handling, and the associated risks of damage and theft
- disruption of cash flow because the project's end date is extended, deferring the release of retentions and securities
- subcontractors are delayed resulting in them claiming delay costs

Many of these costs are difficult to demonstrate and prove to the client. Frequently, clients don't understand the consequences of their actions. The contractor should do whatever they can to ensure that the client delivers access and information in accordance with the schedule requirements so that the project isn't delayed.

Acceleration

Unfortunately, contractors often end up placing more resources on a project than were allowed for at tender stage. This is often due to:

- the scope of works increasing
- the client providing access, drawings or information late
- the client requiring earlier access or completion
- the contractor being delayed by the client's contractors or workers

Often the client forces the contractor into accelerating the work by not accepting the contractor's revised schedule taking these factors into account. This is obviously unacceptable, and while the contractor has some obligation to try and accommodate the client, the client has an obligation to accept responsibility for changes and delays resulting from their actions, and to compensate the contractor for the costs incurred in mitigating delays, and achieving earlier completion dates.

It's preferable that any acceleration claim is agreed with the client before the contractor incurs the additional costs of acceleration.

The contractor needs to put the delays onto the approved contract schedule and demonstrate their entitlement to an extension of time. Once the entitlement is established the contractor can investigate shortening the schedule to comply with the client's revised information and dates. In doing so the contractor must establish what impact this new acceleration schedule has on the works, and what additional resources are required to achieve the new schedule.

This new acceleration schedule must be formally submitted to the client together with the claim for acceleration.

Should the client fail to acknowledge the claim, still insisting that the contractor complies with the original contract schedule, then the contractor must notify the client that they are proceeding with constructive acceleration and that they expect to be paid compensation for this in terms of their claim.

Logging variations

Variations should be logged in a register so that:

- the contractor tracks and ensures that they have priced them
- the contractor follows up ensuring that the client has approved them
- the contractor carries out the work (variations are sometimes overlooked or forgotten resulting in the contractor having to do them after they've completed the contract, or in the worst case, even having to redo completed work to comply with the variation request)
- the contractor is reminded to claim for the work

Getting paid for the variation

It is one thing to submit a variation, but it also needs to be approved and paid. Often clients are slow to approve claims, and the process for payment may take even longer, particularly if the project budget has been exceeded.

Normally, once a client has accepted a variation they have to issue a contract amendment or change order, and contractors are usually not paid for the variation until one is issued.

Many contractors carry out additional work in good faith, in accordance with a site instruction, but then it's either not paid, or isn't paid in full, because the client disputes the costs that the contractor has claimed.

Project Managers should track variations, and ensure change orders are issued. Often I've waited several months for these to be processed.

Change orders must be in writing and should:

- have the correct variation value
- have conditions which are acceptable to the contractor
- include acceptable completion dates
- be signed by both parties
- have a clearly defined scope
- refer to the terms of the original contract

Failure to check these details may result in the contractor inadvertently agreeing to conditions or milestones which aren't acceptable, or having a change order which isn't valid.

Don't exceed the value of the order or variation

This sounds fairly simple and yet it's a common mistake. Project Managers must be aware of the value of an order or variation and constantly track and update the value of work done to ensure that it's not exceeded. The client is in their rights to not pay for work completed in excess of the order. Sometimes the client might not have the funds available to pay for work in excess of the order. Even if they agree to pay the additional value, they will be unhappy that the value was exceeded and they were given no prior warning of this. Furthermore, payment may be delayed while the client amends the order.

The Project Manager should timeously warn the client if the value will be exceeded so that the order can be amended. Often clients take several weeks to issue a revised order. The contractor may be instructed to cease work (in any event it would be prudent for the contractor to stop and wait for the amendment as there's always a risk the client won't agree to pay for further work) while they await the revised order, resulting in the contractor's resources standing idle.

Re-measurable contracts

When finalising re-measurable contracts based on the client's bill of quantities ensure that:

- the actual quantities are measured accurately
- the description of the item or task hasn't changed
- the specifications are the same

Check the overheads and ensure:

- that they are claimed in full
- depending on the terms of the contract, that they are adjusted if the final quantities, value or the time period are varied

Day-works, cost recovery and cost reimbursable contracts

These are contracts, or variations, where the client has agreed to pay the contractor for their actual costs plus a mark-up, which includes the contractor's profit and overheads. Sometimes the client agrees to pay the contractor a set rate for every hour that the contractor's personnel or equipment work on the task.

Before undertaking work on this basis it's important to agree:

- the basis of the charges, what the client will pay for and what's included in the rates
- how the contractor will be reimbursed for unproductive time (for example moving personnel and equipment between tasks)
- how the contractor will be recompensed for their site overheads, facilities, supervision and management
- who from the client has the authority to agree and sign for the hours worked
- what the client will be providing
- what the contractor's mark-up is
- the type of records and proof required by the client to substantiate the costs
- if there's an overall maximum value that shouldn't be exceeded

It is important that accurate records are kept of all costs and hours worked. These costs may include amongst others:

- transport of personnel, materials and equipment
- handling and off-loading of materials
- taxes
- wastage and breakages
- insurances
- royalties
- mobilisations and inductions
- safety equipment and clothing

Hourly records of work performed should be signed daily by the client's representative.

Staff undertaking the work must understand how the client will be paying for the work and what items they should be recording.

Records should agree with the numbers recorded in the daily diaries and weekly reports to prevent confusion, and the possibility the contractor is reimbursed for a lower number of people.

If there is a maximum value for the contract or variation order it should not be exceeded and the client should be advised well before this figure is reached.

Punch lists

Clients often include items on their punch lists which are:

- not included in the original scope or specifications
- damages caused to the facility by the client's employees or their subcontractors

- normal maintenance items which should be for the client's account

The contractor should advise the client that these items constitute a variation and not attend to them unless they have been issued with a variation.

Escalation and rise-and-fall calculations

Some contracts allow the contractor to claim escalation, or rise-and-fall, on the value of work done. (This subject is discussed in Chapter 3.) I've seen many projects neglect to claim for the rise-and-fall, or they calculate the amount incorrectly.

Some escalation formulae can be complicated, so ensure that they are applied properly. Also, check that the figures used in the calculations are correct, applicable to the region and are the ones nominated in the contract.

Dealing with cost increases

With long duration projects, costs will inevitably increase. Ideally these increases should have been allowed for in the tender calculations. However, when preparing variations, particularly those of large value, it's worth considering the cost increases, and where possible adding them into the variation.

Additional work

While working on a project it's often possible to obtain additional work, either from the same client or another one that's working on the site. If this work can be carried out at the same time as the original project there's potential to share resources between projects which improves utilisation.

I have often constructed projects, then remained on site to construct further phases, which have been profitable for the company.

Sometimes, we have provided services to other contractors on a project, for instance selling concrete from our concrete mixing plant or hiring out equipment like cranes. Naturally, care should be taken that this doesn't interfere with the progress of the main project. In addition, ensure that there is a purchase order in place which has acceptable payment conditions and takes into account items such as insurance and liability. We've had instances where we have struggled to be paid for work we undertook for smaller contractors, and in a few cases, even righting off the debt.

Alternate sources of income

Sometimes there are alternate avenues of revenue on a project. One of these is to advertise on hoardings, scaffolds and even on tower cranes. Normally permission must be sought from clients and local authorities. Some advertising can be lucrative, however, care should be taken that the adverts don't appear tacky and diminish the professionalism of the contractor.

Again, these alternate sources of revenue must not detract from the primary goal, which is to complete the project successfully.

Investment income

If the company is operating profitably, with a positive cash flow, it should have cash in the bank. It's important that this cash is invested where it can receive the maximum revenue, but where it's readily available should the company require it to cover operational expenses. Some contractors make a large portion of their profit by managing their cash flow carefully and earning additional income from the invested funds.

To be able to invest surplus funds to earn the best return it's essential the company has accurate budgets and cash flow projections.

Contract bonuses

Sometimes clients are willing to pay bonuses for early completion. Even if these aren't offered in the tender it could be something which is negotiated with the client if it's known that they are desperate to move into the facility ahead of schedule.

Payment of subcontractors and suppliers

Payment of subcontractors and suppliers should be done in accordance with the agreed terms and conditions. Payments should:

- be made on time so that discounts can be claimed
- have the relevant retentions withheld from them
- not be released without completion and receipt of:
    - warranties and guarantees
    - a signed final account
    - the completed punch lists
    - the required spare parts and manuals
    - all commissioning data
- be checked to verify:
    - the subcontractor has completed the work they've claimed
    - the work complies with the specifications and quality requirements
    - all materials have been received undamaged
- only include claims and variations which have been agreed with the contractor and where necessary the client
- only be made for unfixed materials if this is in accordance with the subcontract agreement and providing the required guarantees or sessions have been received

Ideally the subcontractor shouldn't be paid more than the contractor has claimed from the client.

Supplier payment terms

By negotiating better supplier payment terms the contractor may be able to pay suppliers later, which helps the contractor's cash flow, making the company more profitable.

Alternatively it may be possible to negotiate a discount if the payment is made earlier.

Back-charges and services supplied to subcontractors

While it shouldn't be the contractor's aim to make a profit by deducting money from subcontractors, it's often necessary to back-charge a subcontractor for costs the contractor incurred on behalf of the subcontractor. These costs include:

- services supplied to the subcontractor which they should have allowed for and provided such as:
  - accommodation
  - transport
  - clearing their rubbish
  - access scaffolding
  - providing lifting equipment and offloading facilities
  - provision of water or power
  - offices and other site facilities
  - plant and equipment for their work
  - personnel
- repairing work damaged by the subcontractor
- repairing the subcontractor's defective or poor workmanship

Sometimes these costs can be significant. Subcontractors should be timeously warned that they will be charged for services and what the costs are. When necessary, they should be given sufficient written notice to rectify problems. Where possible, time sheets for equipment supplied should be signed daily, and the back-charges should be agreed and invoiced monthly so that the subcontractor is fully aware of their liability.

Failure to notify the subcontractor, or get them to sign for the services received, often results in disputes, even ending with the contractor having to withdraw their invoices.

Insurance claims

Certain events and damages are covered by the contractor's or the client's insurance policy. The Project Manager must be aware of what these events are, what the excesses or deductibles are, and the process to follow to claim against the policy. Failure to notify the insurer timeously, or to record and report the event, could render the claim invalid.

Furthermore the contractor must ensure that they have taken all mitigating steps to prepare for and prevent an event from occurring, otherwise they may find that their claim is repudiated by the insurer. For instance, if a vehicle is in an accident and it was found the driver didn't have a valid licence, the vehicle wasn't properly maintained or was un-roadworthy, the insurer probably won't pay for damages. The same applies to items damaged in a flood where the contractor didn't take proper precautions to prevent the works from being flooded.

Equally important is to ensure that the insurer is given accurate information when the insurance policy is purchased and that they are advised of any changes in the project. Failure to do this could render the policy null and void.

Insurance claims should include all the costs associated with carrying out the repairs and making good damage including:

- the material costs including transport, insurance and handling
- all labour costs
- costs of cleaning, removing and disposing of debris
- protection of the undamaged work which is affected by the repair work
- demolition of damaged structures
- supervision and overhead costs
- damage to plant and equipment
- temporary support or access structures
- subcontractors' costs associated with the damage

Unfortunately most insurance policies don't cover for consequential damages. Any delays caused by the event, including their resultant costs, won't be covered.

Negotiate better payment terms with the client

By negotiating better payment terms with the client the contractor can improve their cash flow which saves on interest charges.

Refer to Cash flow in Chapter 8.

These terms should be included as part of the tender submission, although some may be negotiated after the awarding of the contract.

Recovery of GST, VAT, duties and other taxes

The contractor should obtain specific advice regarding the recovery of GST and other taxes and duties paid because legislation varies between countries and states and can change over time.

It's important that all receipts, invoices, and other paperwork are kept so that company administrators are able to submit them in order to recover monies owed. These documents usually have to be tax receipts and in some cases may require specific information to be included on them.

The flip side to duties and taxes is to ensure that the client is charged for all duties and taxes that are for their account. For example, when items are imported there may be import duties and clearing charges which the contractor often neglects to add to their price.

Ensure that the client pays

As discussed previously one of the biggest risks to most contractors is not getting paid for their work. The contractor may have done everything in their power to ensure that they are paid, such as submitting invoices on time and having all day-works, site instructions and variations signed by the client. Unfortunately after all of this, and even despite the good intentions of the client, problems occur and clients experience financial difficulties. It's therefore important to pay close attention to any rumours or news reports regarding the client's financial wellbeing, and to investigate these as soon as possible. If necessary, call a meeting with the client to discuss the reports that you've heard and to understand whether they have sufficient money to pay for the project.

Since the value of work carried out increases every day, the longer the contractor works, the bigger the risk becomes, and the greater the amount of money that's owed. Unfortunately, the contractor cannot just stop work, because this may put them in breach of contract, which would give the client cause to terminate the contract and not pay them at all.

When there's a risk the client will be unable to pay, the contractor should:

- when the valuation payment is late immediately formally notify the client they are in breach of contract
- follow the requirements in the contract for termination, ensuring all notifications are addressed correctly
- delay major material deliveries and purchases where possible
- delay work where possible, providing this won't jeopardise the contractor's rights under the contract
- delay mobilising further equipment or people to site where possible
- remove equipment which isn't essential to the works
- check the validity of payment guarantees
- take precautions to limit the effects of the non-payment on their cash flow
- if possible negotiate that the client pays at least a portion of the outstanding money
- if the client is unable to pay the outstanding money, and the contractor has clearly followed all due process, then the contract should be terminated

Summary

The contractor should maximise their revenue by:

- ensuring that monthly valuation claims are submitted timeously and in the correct format
- maximising the amount claimed in each valuation
- claiming all the variations that they are entitled to
- ensuring that variations are submitted in accordance with the contract and include all additional costs incurred by the contractor
- recording, reporting and claiming for all delays caused by the client or their contractors
- claiming for the costs incurred to accelerate the works
- finding additional work on the project
- ensuring that damages caused by an insurable event are correctly processed and claimed
- investing surplus cash to maximise income, while retaining sufficient funds to cover the company's operating requirements

- paying subcontractors and suppliers on time, in accordance with the conditions in the order, and ensuring that they have completed all the work claimed
- ensuring that subcontractors are correctly charged for services supplied to them
- recovering taxes and duties
- negotiating better payment conditions with the client
- ensuring that the client pays all claims due, on time

Chapter 8 – Financial Management

Many small companies literally live from hand-to-mouth, receiving money from one project and spending it on another, with no idea how profitable they are, or whether they have made money on a project or not. Their only clue to their overall profitability is their bank account, but unfortunately, some owners mix their personal and business accounts, so even this is a poor barometer. The poor control of finances sometimes means that contractors continue to undertake projects which aren't profitable. In addition, they're often unable to claim back monies they're entitled to from the tax authorities.

Larger companies generally have sound financial systems in place which are easy to operate and can give cost updates in real time for every project. Unfortunately, every system is only as good as the operators and the quality of the information that goes into the system. There are some systems that are complex and provide lots of detailed information that's not used by most contractors.

Some construction companies use systems that are more appropriate for other industries, while others use systems which were more appropriate to a time when they were smaller, and they've outgrown them now.

Even when there are adequate financial systems in place many business owners and managers don't understand some of the basic accounting principles which cause them to make inappropriate choices or expensive mistakes. It's therefore important that business owners and managers understand the systems the company is using and that they also have a basic understanding of accounting principles.

Just as important is to ensure that the system is operated correctly across all divisions and facets of the company.

Contractors need to be disciplined in their approach to project and company finances. Deviations from set procedures can quickly result in uncertainty amongst staff, financial mismanagement, loss of financial control and even bankruptcy.

I would also always urge contractors to be financially conservative since the contracting environment is continually changing with unexpected events happening when least expected, such as clients being unable to pay, contracts being cancelled, accidents, disruptive weather events or a shortage of work. Contractors who have stretched their financial resources might be unable to cope with these challenges.

Cost controls and reporting

Cost reporting and cost controls for individual projects are important since:

- they provide an indication as to what types of project are profitable and should be targeted by the company

- the information can be used to improve the accuracy of tenders
- they assist the company with their financial forecasts and budgets
- when a loss is detected it may be possible to prevent further losses and even to recover some of the money
- it makes project staff accountable

The reporting systems must:

- be simple to operate
- provide information that can easily be analysed
- be compatible with other financial and operating systems the company uses
- be adaptable to use on small and large projects, as well as different types of projects, so that one common system can be used throughout the company
- if possible tie-in with the information provided from the tender process
- provide reliable information
- provide information that is current

Not only should companies have a good cost system, but equally importantly, they must ensure that the information produced is analysed and used to improve the company and project performance and operations.

Cost to completion

Cost to completion is a form of project cost reporting. In this method an estimate is made of all the costs that are expected to be incurred for the remainder of the project and these are added to the costs already expended. The total costs are then compared with the expected final revenue.

To make the results more accurate it's possible to split the costs into different sections like labour, materials and equipment, or even more detailed into subsections of different materials.

The reliability of this method of cost reporting depends on the accuracy of the data for the costs incurred to date and also requires that accurate forecasts are made for the future costs. It goes without saying that the expected final revenue should also be accurate and take into account any deductions or reductions in revenue.

Cost versus allowable

Another method of cost reporting is to compare the actual costs incurred to date on the project with the revenue earned so far. This can be done for the project as a whole, for individual tasks or for different cost codes such as labour, plant and material.

The accuracy of this method depends on the correctness of the calculations of the costs as well as the accuracy of the revenue against which the costs are compared.

Cost reporting using the schedule

If the schedule has been correctly resourced it's possible to calculate the resources that should have been expended according to the schedule and compare these with the actual resources expended. This can usually be done for manpower and equipment.

Cost codes

To enable accurate cost reports to be prepared codes should be allocated to the different tasks or trades. These cost codes should:

- be simple and easy to use
- be uniform across all projects
- have a prefix unique to each project
- allow for expansion for future new items, materials and activities
- be allocated from the start of all projects
- be used by everyone associated with the project
- be understood by all personnel using and allocating them
- tie-in with the codes used in the tenders

If the costs and the revenue of the different tasks and items can be correctly allocated to a cost code, it's possible to get an accurate portrait of whether that task, or item, is making or losing money, and appropriate steps can then be taken to rectify losses.

Investigate losses – don't cover them up

When a project loses money it's important to establish the reasons for the loss so that further losses can be prevented, and if possible, so the losses can be recovered. There are many reasons for a loss including:

- wastage
- poor productivity (Refer to Chapter 6)
- poorly planned and organised projects
- defective materials
- theft
- poor quality work resulting in rework
- poor performance of subcontractors
- failing to claim for work done
- changes to the working conditions caused by the client such as:
    - delayed access
    - increased scope
    - variations and changes
    - the late issuing of drawings and information
    - changes to the schedule
    - restrictions on work hours
    - changes in specifications or tests required
    - client-supplied services not being available or not in the required quantities

- site conditions different to those tendered for such as:
    - the presence of rock
    - other contractors impacting on the work
    - the client's activities impacting the works
    - the presence of hazardous materials
    - available borrow pits being further from the work site or more difficult to access
    - the material having different properties than specified, making it more difficult to work with
- unusual adverse weather conditions
- incorrect tender assumptions or calculations

If the losses are due to the client, or to changed conditions, it should be possible to submit variations for the additional costs. If the loss is due to the contractor then management needs to identify steps to mitigate and recover the losses. Sometimes the loss is due to mistakes in the tender and there's little the project team can do. However, it's important that the estimating team is made aware of the errors so they don't occur again.

Project Budget

Each project should have an accurate budget. The purpose of a budget is to provide an estimate of the profit or loss the project will make. The budget should be as accurate as possible and should be updated during the course of the contract to take into account any changes in circumstances or errors in the assumptions made when the budget was first prepared.

Case study:

I started work with a new company and was given a project to manage which had been running for two months. The budget for the project was broken down into material, labour, equipment and subcontractor costs. Referring back to the tender I couldn't find how the numbers were derived. After asking several people I finally found out that the manager had looked at the contract sum and simply proportioned the various cost components, as he thought best. There was no science in what he had done, and yet for the rest of the project every month he would compare the cost report with the budget and tell me we were making money on materials but losing money on labour. A completely nonsensical and pointless exercise!

The budget is usually derived from the estimate for the tender. Ideally the Estimator has accurately calculated the cost and quantities of materials, labour, equipment and subcontractors required for each activity. In addition they should have calculated the cost of management, supervision and overheads to manage the project, and to all of this added profit and overheads. Using this as a basis the Project Manager should check that these costs and assumptions are valid and revise them to take into account any changed circumstances.

When the Project Manager plans and starts the project they may find:

- that the Estimator has made errors, either over or under-estimating the costs
- they are able to procure materials more cheaply
- they may decide on a different methodology which may for instance reduce the labour costs but increase the material costs
- the client may have added to, or reduced, the size of the project compared with the tendered project
- by changing the methodology the overall duration of the project may change which affects the management, supervision and overhead costs
- equipment that the Estimator thought was available is no longer available, resulting in more expensive methods and equipment being used

Using a proper and accurate budget the project can obtain an accurate measure of whether the project is making or losing money.

Of course preparing a budget shouldn't only be an exercise to find negative errors and to downgrade the forecast profit, but it should also look at potential upsides so that it's an accurate forecast. Many of the budgets I prepared for my projects showed we would achieve a better profit than the tender – and in most cases we even exceeded these profit expectations.

Company Budget

From the various project budgets the company should prepare an overall budget which would also include an estimate of all the company's running costs and overheads which aren't covered by the individual projects. An accurate budget is necessary:

- to show shareholders or owners what profits or losses to expect
- if the company requires additional finance from lenders
- so that owners and shareholders aren't paid out money which may be needed later to finance the company
- to plan purchases of equipment
- to ensure that there are sufficient funds to pay loans and liabilities
- to ensure that the company's cash flow remains positive
- to set a target for additional work the company must win
- to measure the performance of the company

Levels of authority and financial controls

It's important that companies have proper financial controls in place as well as suitable levels of authority to make purchases and payments. Without them people are free to spend money as they decide and there's a risk:

- items which aren't required are purchased
- money is stolen
- the company has insufficient funds available to cover operating expenses
- the company purchases items which they cannot afford
- payments are duplicated

- mistakes are made when people who don't have the experience or knowledge are allowed to make or authorise payments

Most companies should have a clear policy of what individuals are allowed to sign or agree. Some of these would include:

- payment to subcontractors and suppliers
- payment of wages and salaries
- placing purchase and subcontractor orders
- submitting tenders
- submitting variations
- signing contracts
- authorising time sheets
- making capital purchases

Different levels of authorisations are usually put in place which depends on:

- the magnitude of the cost
- what the item is
- the frequency of authorisation
- the level of trust with individuals
- the size of the company
- the person's position in the company

Sometimes, with items of large value there may in fact be a whole process of authorisation which needs to happen in sequence. For instance a subcontractor payment may have to be compiled by a Contract Administrator, then checked and authorised by a senior Contract Administrator, who then passes the payment to the Project Manager who checks and signs it before giving it to the Contract Director for authorisation, and sometimes, with particularly large payments, the Divisional Manager may have to finally authorise the payment. These multiple authorisations can be cumbersome and take time, but when large payments are made they are particularly important to ensure that mistakes aren't made with the payment. Of course they are only effective if each individual actually checks the payment, and I've had cases where incorrect calculations have been made and these weren't picked up despite the payments being authorised by three or four different individuals.

Sometimes the authorisation process can make it difficult to operate a business. This is particularly the case in small companies where the owner has to authorise all payments. Often the owner is out of the office trying to secure more work, resolving problems on a project or away on leave. When they cannot be found to authorise a payment it may mean a supplier or subcontractor isn't paid on time, or even that wages and salaries aren't paid. In these cases it's important that there are other individuals who are more readily available delegated to the task.

Of course it doesn't help the process if projects don't submit their paperwork on time and send payments through at the last minute. As a Project Director I found it frustrating when payments were sent to me for authorisation on the afternoon before they were due to be paid (or even sometimes on the day payment was due), since I wasn't necessarily always in the office.

It's also important that when the person who has to authorise payments is going to be unavailable for an extended period, or over the period when payments normally need to be processed, they delegate other suitable people to authorise on their behalf.

Checks and controls

In this day and age of theft and fraud it's important to have sufficient checks and controls in the operating systems to pick up and, more importantly, to deter fraud.

These checks should include ensuring that:

- the item has been received and complies with the quality requirements and the specifications
- the work has been carried out in accordance with the order including supplying all quality documentation, spares, and warranties and completing all commissioning
- the value invoiced does not exceed the value on the order
- deductions have been taken into account
- the agreed discounts have been taken
- retention is withheld where applicable
- the correct amount of tax is added
- the invoice hasn't been previously paid
- there aren't any arithmetic errors

To facilitate some of the above it may be necessary for the appropriate people who have the required knowledge to check and confirm the item has been received and complies with the requirements of the order.

It's also essential that the payment process is clear so that it's possible to follow what the payment is for and what deductions have been made and why they were made. If it's not clear it's possible that when future payments are made the supplier receives more money than is due, or deductions are accidently reversed.

To enable this checking process it's important to tie-up the order, invoice, delivery, batch and payment numbers. Equally important is to ensure documents are filed and stored in the correct place and sequence and are readily available for several years in case disputes with suppliers arise later.

Documentation

It's important that all documents and invoices are kept and filed so that the accountants can accurately calculate all expenses associated with running the company. Some small company owners mix their personal expenses with the company expenses, which either means they declare less expenses than they actually incurred, or, they end up declaring expenses which weren't legitimate which could expose them to being penalised when tax authorities conduct an audit.

All documentation should be kept for a minimum of five years. Some countries have laws requiring documentation to be kept longer (for instance in Australia it's seven years), and I would certainly advise that it's kept as long as possible. The documentation should be kept in an orderly system where it can easily be accessed when required. The storage areas must be weather-tight, and be checked for the presence of rodents and insects.

Case study:

I've had occasion where I've had to look for documentation that was stored in a stifling hot sea-container. The documentation was all jumbled making it almost impossible to locate what was required. The documents became more mixed-up every time someone looked for another document. Furthermore, the container wasn't waterproof and leaked when it rained resulting in much of the documentation being illegible and useless.

Computer records must be safely stored in a location where they can't be damaged by fire or heat.

There are specialist companies who provide safe document storage, although it is advisable to inspect these premises to ensure they're what they are promised to be.

Fraud

Fraud is a major problem faced by most companies, much of it carried out by employees of the company. Unfortunately most fraud remains undetected, or if it's detected it's usually too late and little of the stolen money is recovered. The best way to prevent fraud is to employ trustworthy individuals, but even with the best reference checks and employment procedures it's impossible to guarantee this. Even the most trusted employees can be tempted to commit fraud when presented with an opportunity to make easy money, or if they become addicted to drugs or gambling, or their home and financial circumstances change causing them to resort to stealing.

Some types of fraud include:

- misuse of company property such as vehicles and telephones
- charging the company for fuel or other purchases which were purported to be for company use and weren't, or charging for travel costs which weren't company related
- using company materials and equipment for their private work (like building or renovating their home)
- misusing a company credit card
- falsifying time sheets by adding additional imaginary people, or adding extra hours for employees
- asking suppliers or subcontractors to inflate their invoices on condition that a portion of this is paid to the employee
- hiring of company-owned equipment to a third party and pocketing the payments for the item

Case study1:

I've known a director that built houses using people and materials from their projects, all paid for by the company. When it appeared this practice was condoned it made it difficult for the rest of us to ensure that others in the company didn't do the same thing because they saw the practice was accepted.

Case study 2:

The name of our company consisted of four initials. The person responsible for receiving and processing payments made to the company opened a bank account for a company they created with a name consisting of four words which began with the same letters of our company name. When a cheque was received in our company's name, instead of banking the cheque in our account she deposited the cheque in the account she created with the similar name. Sometimes when a client phoned to ask for our company's bank details so they could transfer a payment directly into our account, she gave them the details of the account she had created.

The problem was only detected because we had systems in place which picked up the discrepancy between the actual money paid into our account with the receipts issued. However despite her going to prison not all the embezzled money was recovered.

Case study 3:

We used to rent houses for our employees to stay in while they were based on a project. On one particular project we rented a property for our Project Manager and his family. Unbeknown to us, he sent his family back to their family home and he moved into one of the houses used for single accommodation. He terminated the lease on the property rented for him and his family. He didn't inform our Head Office that the lease was terminated, but, instead he pretended that the person letting the property had changed their bank account and gave our accounts department details of a new account where the monthly rent should be sent. Fortunately our accountant spotted that the new account was in fact the same account into which his salary was paid. The scheme ended before it had begun. However, we'll never know where else he defrauded the company.

The best way to prevent fraud is to ensure there are systems in place which remove the temptation to steal and that prevent theft from occurring. There should also be sufficient checks and controls in place to detect fraud when it does occur so it can be stopped.

Sometimes the fraud is perpetrated by people outside the company. Some of the more common types of fraud would be:

- hacking into the company's internet bank account
- paying with fraudulent cheques
- issuing of guarantees which don't have the correct company name, contract sum, date or which may in fact be fake which are then worthless should the company need to claim against them
- stealing cheques made out to the company
- invoicing for materials not delivered or for work which wasn't carried out
- supplying substandard materials
- double invoicing for the same item
- not delivering the full quantity shown on the delivery note

- charging for repairs and spare parts which weren't required
- charging exorbitant fees for basic services

Apart from fraud there is the risk of common theft which is covered in Chapter 11.

Cash flow

Negative cash flow probably causes more construction businesses to run into financial trouble (leading to their closure) than any other cause. Even a profitable project can cause a company financial problems if the cash flow is negative.

Negative cash flow is when the contractor is paying money to suppliers, equipment hire companies and subcontractors or in wages and salaries before the client has paid for the work that has been completed.

Unfortunately most construction projects are usually cash negative to some extent. Many clients hold 10% cash retention until the end of the project when this is reduced by half. This means that if the project is tendered at anything less than a 10% profit the project is usually automatically cash negative until at least the end of the project.

In addition, most clients only pay the contractor thirty days after the contractor submits an invoice. These invoices are normally submitted at the end of each month. Many contractors pay their workers fortnightly or in some cases weekly. This means as a minimum the contractor has paid out up to seven weeks of wages before the client pays for the work that these personnel have completed. Smaller contractors sometimes have to pay suppliers before they will release materials.

There are, however, a number of other factors that make the cash flow situation even worse. Many projects have payment terms longer than thirty days. In addition some clients habitually pay progress claims late or not in full. Of course, the ultimate knockout blow for many contractors is when clients don't pay at all. This could be a result of the client disputing the value of work, defaulting on the contract or going into liquidation.

Yet, even with the odds stacked against contractors, they often make their cash flow situations worse by submitting their progress valuations late, accepting payments late or not claiming fully for completed work.

In my book on project management I cite an example of one company that had completed over ten million dollars of work on a cost recovery basis but didn't have any of it signed off and approved. They had invoiced for only five-hundred thousand dollars of the work. This was obviously very stupid and was wrecking their cash flow. Most companies couldn't carry this negative cash flow. Yet, many companies don't claim all the work they've done in the month.

To improve the cash flow situation contractors must submit their monthly progress valuations on or before the due date. Some clients only run progress payments on a particular day in the week, or month, so missing a submission date could cause the client to delay payment by up to a month. The valuations must be submitted in the required format and with the required supporting documentation since many clients will use any excuse to delay or reject a monthly claim. It's important to track the progress of the payment through the client's payment system. With major clients there may be several people that check and approve the valuation and payment. Sometimes, the process is disrupted when someone is absent, or the claim simply gets 'lost'. I've had more than one client who consistently paid progress claims late, always with some excuse about our valuation being late, the claim being incorrect (either arithmetic errors, insufficient supporting documentation or disagreement with our progress) and people in the approval process being absent. Of course many of these problems were only reported to us when the payment was due, despite the client having had the claim for thirty days.

It's also important to ensure that all the work done is claimed in the valuation.

Often the monthly valuation is due, say, on the twenty-fifth of the month which normally means the contractor prepares it on about the twenty-third, while the valuation is actually for the full month. This means that about a week's worth of work is carried out after the claim is prepared. Often clients will accept that a forecast of the expected work in the last week is included in the valuation. Obviously the contractor has to be reasonably sure that this work will be completed because if it isn't, and the claim exceeds the actual work done, there's a possibility the client will reject the valuation in its entirety, which would delay the payment.

With some clients it's possible to add a few extra days of progress into the valuation because by the time they actually review the claim (which may be several days after the valuation was submitted) the extra work has been completed.

Some contracts are structured such that payments are only made when the contractor achieves particular milestones. It's important that Project Managers understand what these milestones entail and ensure they are met. It's obviously pointless to achieve 99% completion if payment is only made for 100%. Often contractors take several weeks to complete the last few items (which may just be completing documentation), which delays payment.

As mentioned, retention monies held by the client impact a contractors' cash flow considerably, and yet, many contractors fail to recover them as soon as they can. For retention to be released contractors must fulfil certain obligations in terms of the contract which includes handing over the works as well as submitting various documents (such as quality documents, as-built drawings and guarantees) and completing all punch list items. It's vital that these obligations are met as soon as possible to enable the release of the retention, as well as to enable the warranty period to commence which will culminate in the release of the final retention. It's then important to track the warranty period, and timeously request that the client undertake the final inspection at the end of this period, thus enabling the release of the final retention.

Many clients don't pay for materials which aren't fixed in place on the project. Clearly these materials (particularly large value items) should be kept to a minimum and only delivered to the project when they're required to be built in. This is sometimes difficult to control, and if there are delays with the delivery it can result in the project being delayed with costs which far outweigh the benefit to the cash flow of receiving the materials 'just-in-time'. Where possible contractors should build in large expensive items before the monthly claim is submitted. Deliveries can also be planned so that they arrive at the start of the month rather than the end of the preceding month.

Contractors can do many things at the time of preparing the tender to improve their cash flow such as:

- checking that the client can meet their financial obligations on the project
- researching whether the client is known to pay the full value of the monthly valuations on time
- negotiating more favourable payment terms which may include:
  - paying the valuation within a shorter period than stipulated in the tender
  - being able to submit invoices at an earlier date or more frequently
  - the client withholding less retention
  - payment for unfixed materials on site
  - reducing the amount of retention withheld
  - replacing the retention money with a surety bond
  - asking the client to provide a payment guarantee which could ensure that if they got into financial difficulties the contractor could claim against it
  - requesting the client to make an advance payment (particularly to cover the purchase of major items of equipment or material)
  - structuring the tender in a manner that a larger portion of the project overheads are paid at the start of the contract, or that work done earlier in the project has a higher value than work done later (I've run many contracts with over-claims running in excess of a million dollars just by structuring the tender correctly and ensuring that our monthly progress valuations were maximised)
- requesting the client release retention money earlier by:
  - shortening the duration it's held
  - releasing it in tranches as milestones are achieved
- reducing the value of the guarantees
- allowing the guarantees to be released earlier or when the important milestones have been achieved

Variation work often negatively affects the cash flow.

Case study:

One of my contracts had a tender value of eight-hundred thousand dollars. However, in the course of the project we submitted numerous variations because the scope increased. We also encountered rock which wasn't allowed for, and the client consistently delayed the project because they provided access and information late. We submitted all variations timeously during the course of the project and eventually the value of the contract was nearly two million dollars. Our client, nonetheless, consistently refused to deal with these claims, which meant they didn't issue the required variation orders and consequently we weren't paid for these variations. Only after we took legal action did our client eventually agree to consider our claims. We finally agreed on a final contract value of one million six-hundred thousand dollars which was double the original contract value. The delay by the client in agreeing the value of the variations meant we were paid nearly six months after most of the work was completed – which had a crippling effect on our cash flow.

Even with a more responsive client the contractor can wait several months to be paid a variation since many clients have to go through a process to approve them and issue the amended order. In some cases, the variation may result in the client's approved contract value being exceeded, which may require them having to apply for additional project finance – causing further delays. It's therefore preferable that contractors don't undertake variation work until a variation order is in place. To pre-empt problems the contractor must notify the client immediately a variation is noticed. These variations should be priced immediately and the client urged to agree a value and issue the appropriate variation order as soon as possible.

When undertaking work which is re-measurable, if the contractor isn't vigilant, they may find that the contract is complete and only when doing the measurements and settling the final account realise that the quantities have exceeded the tender quantities resulting in the contract value being exceeded. Therefore, contractors should ensure drawings are measured as soon as they are received and the client timeously advised if there's a chance the contract value will be exceeded.

Payment terms of subcontractors and suppliers also affect cash flow. Where possible, contractors should negotiate terms of thirty days or more. However, whatever happens, the contractor must ensure suppliers and subcontractors are paid in accordance with the conditions of their respective orders and contracts. Late payments not only result in the contractor losing payment discounts, but can also lead to a breach of contract. In addition, once a contractor gets a reputation of paying late it's difficult to negotiate more favourable payment and financial terms with suppliers in the future. It's important for contractors to build a reputation with their suppliers and subcontractors of paying fairly and on time.

Don't trade on over-claims – but do over-claim

Where possible the contractor should claim for as much work as possible in the interim valuations as it's always better to have money in your bank account than in the client's and it improves the project's cash flow. However with any over-claim it's important that when the cost report is compiled the revenue in the report is adjusted to remove the effects of the over-claim so that the cost report doesn't produce a false result.

Case study 1:

One company I worked for had a number of projects that were showing healthy profits. However, as the projects progressed, these profits evaporated, with many turning into losses. So what was the problem? Well, the cost reports had merely compared the costs with the monthly valuations. These valuations were grossly inflated, yet, no adjustments were made in the reports for these over-claims. The dramatic reversal in fortunes was because towards the projects' end the client's valuations were less than the costs incurred because the Project Manager had previously already claimed revenue for work which wasn't yet executed.

Case study 2:

Another project I worked on was not achieving the budgeted profit so the Managing Director insisted that we factor in additional revenue from claims we would lodge against the client and the subcontractors. This obviously caused problems when some of these claims didn't eventuate.

It's dangerous to assume the client will pay for a variation or claim because:

- the client may not approve the variation
- the client may find errors in the variation and only pay part of the claim

Tax

Unfortunately all companies have to pay tax if they make a profit (unless based in a tax free haven). There are different taxes that have to be paid at different times which vary between regions and countries. Some of these taxes include:

- tax on the company profits
- payroll tax
- GST or VAT
- tax deductions from employees' wages
- local rates or taxes on properties
- special taxes

Failure to pay these taxes on time may result in penalties.

It's best to obtain expert advice so that the company is structured to minimise the amount of tax, to maximise deductions (such as for depreciation), and to claim for subsidies or payments (such as for training and employing apprentices).

Tax can become complicated when working in different states and countries and the company needs to have a good understanding of how the tax regime varies between the countries to ensure the correct tax is paid on time and also that tax isn't paid twice. Some of these countries may even have different tax year ends.

On occasion investment advisors and accountants invent intricate and complicated plans for avoiding tax. Care needs to be taken with these schemes as many aren't legal, while others simply defer the tax liability to another year. These schemes often cost more money than they're worth, and many are risky and nothing but a ploy to defraud the company of money.

Joint ventures

Sometimes two or more contractors form a joint venture, combining their resources to construct a project. For all intents and purposes the joint venture operates as a separate company in which each of the contractors owns a share (which is in accordance with their percentage participation in the joint venture). A new company, in the name of the joint venture, is usually registered for tax purposes with its own bank account. The joint venture is normally run by an executive committee made up of representatives from the different parties and in accordance with the joint venture agreement.

The committee meets monthly to review progress, cost reports and bank statements, and to take decisions regarding capital expenditure, investment of surplus funds, the payment of profits and the distribution of assets at the end of the project.

Each company invoices the joint venture for staff and personnel they have seconded to the project as well as for their equipment hired to the joint venture. These invoiced amounts should either be at predetermined and agreed rates or at their proven costs.

The joint venture normally places orders, in the joint venture's name, with subcontractors and suppliers, and then pays for the work done.

The joint venture will be audited and be required to pay taxes as any other company would.

Only once the joint venture has settled all its liabilities and the project reached the end of the defects warranty period can the joint venture company be wound up and closed down.

Save for the lean times

Construction is a very cyclical business, experiencing years when there's sufficient work (sometimes even too much), followed by years when there's insufficient work and lower profit margins. It's not only important that contractors are able to adapt to these lean times, but they also need to have sufficient funds to cover their recurring payments as well as to cover their operating cash flow requirements.

I've seen contractors in good times go out and purchase new equipment, offices and vehicles and pay their directors and owners large bonuses and dividends. Unfortunately, these same companies often ran into financial difficulties when the construction market became difficult and they were unable to pay for the extravagant purchases they had made, they had insufficient work for the new equipment or had to reduce their staff complement making the extra space in their new office redundant.

Summary

To be able to operate effectively construction companies need to be financially sound. This not only entails making a profit on all their projects but also means they must manage their cash flow so they have sufficient funds to pay their ongoing expenses.

In order to manage their financial position properly contractors should:

- prepare regular cost reports for each project
- have suitable financial controls and levels of authorisation in place
- implement measures to prevent and detect fraud
- prepare cash flow forecasts for individual projects as well as the company as a whole
- seek ways to improve their cash flow
- find the causes for losses and take action to prevent further losses and to recoup the money lost
- prepare project budgets
- prepare a budget for the company
- get expert advice to minimise tax but always ensuring that their tax affairs are up to date and are within the legislated guidelines
- file documentation correctly and store it where it's easily retrievable and won't be damaged
- take care when setting up joint ventures to ensure that there are proper agreements in place between the partners and that the finances of the joint venture are kept up to date and comply with legislative requirements
- ensure that the company has sufficient funds to cover the lean times

Chapter 9 – Contractual

Larger clients issue the contractor with a contract which is used to govern the project. Unfortunately some contractors don't read these contracts which could result in:

- the contract being heavily biased in favour of the client
- the contractor not performing the work in accordance with the contract
- the contractor binding themselves to contractual conditions which they are unable to fulfil
- the contractor not ensuring that the client fulfils their obligations under the contract

Failure to read, understand and comply with the terms of the contract often results in:

- the contractor failing to claim their entitlements in terms of the contract
- the contractor forfeiting their right to claim
- expensive and lengthy legal disputes

When dealing with smaller clients such as home owners, contractors usually use their own form of contract which they issue to the client. In many countries there are standard forms of contract which can be used for this purpose. When issuing a contract to a client it's important to ensure the contract is:

- appropriate to the type of work (for example it's usually inappropriate and unnecessary for a contractor constructing garden paving to issue the homeowner with a twenty page legal contract)
- clear and unambiguous without contradictory clauses
- is fair and reasonable
- legally binding

When drafting contracts there's always the temptation to over-complicate the contract, or to delete clauses and insert new terms and conditions which may be more onerous to one of the parties. When this happens it can result in clauses being contradictory, even making the contract unenforceable.

It's good practice to obtain appropriate legal advice when drawing up contracts or signing them.

What is a contract?

The contract is an agreement between the client, who contracts to pay for certain work, and the contractor who agrees to perform the work. It's there to protect both parties and sets out their respective rights.

The contract should include the following:

- signatures of the authorised representatives from all parties
- the names and addresses of the contracting parties
- include a scope of works or reference to drawings
- the location of the project
- the date when the contract comes into effect
- the tender sum or value
- milestone dates or a schedule (or reference to a schedule)
- payment terms and conditions
- reference to the tender documents (if these form the basis of the contract)
- include the contractor's exclusions that were made in their tender submission (if these were agreed to by the client in the tender negotiation process)
- the responsibilities of the various parties
- specifications (or refer to specifications)
- the defects and warranty periods
- procedures for submitting variations
- penalties or liquidated damages (their quantum, when they will be applied and, if applicable, a maximum value)
- a termination clause
- dispute resolution procedures
- definitions of the terminology used
- particular conditions applicable to the project site (such as wage agreements, working hours, restrictions on access to site, security, safety, environmental concerns and the use or protection of existing or neighbouring property)

When does a contract exist?

For a contract to exist there must be a formal offer which must be accepted in an unequivocal way. (In other words if a painter submits a quote for painting a house and the homeowner tells the painter he likes the price it doesn't mean the homeowner has accepted the painter's price and appointed them to do the work. For a contract to exist the homeowner would have to say something to the effect that they accept the quote and would like the painter to undertake the work).

The contract can be verbal, which can be as legally binding as a written agreement, though, a verbal agreement is often problematic since it's difficult to prove what was said and agreed.

An offer can be withdrawn before it's accepted provided that the withdrawal is explicit and is preferably in writing. (To use the same example above it won't be good enough if the painter submits the quotation and in later discussion with the homeowner says he's not sure if his price is correct, rather, the painter needs to formally state they've submitted the wrong price and are withdrawing their quotation.) Sometimes clients include clauses in their tender documents which prevent the contractor from withdrawing their price once it has been formally submitted.

In certain cases a contract may not exist if for example:

- one of the parties is; a minor, mentally impaired, bankrupt, a prisoner, or under duress or unfair influence
- the agreement includes false statements
- the contract is fundamentally flawed
- one of the parties knowingly causes the other party to commit an illegal act
- there's a condition precedent (for example the client states they need to find finance for the work and they are unable to do so)
- an act of God makes it impossible to carry out the terms of the contract (for example a builder is contracted to renovate a house but before they start the house burns down)

It is, nevertheless, important to be familiar with the laws governing the contract as these differ between countries, impacting on whether the contract is legally binding.

The legal basis of the contract

The laws governing the contract vary from country to country and aren't always the laws of the country where the project is located. Often clients or managing contractors are based in a different country and elect to use the laws of their home country. The laws and legal systems may be unfamiliar to the contractor and disputes may have to be referred to the judicial system in that country, which will require the contractor to travel there to pursue their claims.

As discussed in Chapter 3 it's preferable during the tender stage to get the client to change the legal basis of the contract to one familiar to the contractor. Failing this it may be appropriate to obtain legal advice from a person familiar with the legal system specified in the contract.

The same contract – but is it?

Contractors often become familiar with particular contract conditions, or repeatedly work for a client using the same contract conditions. It's however important to read each contract carefully, because contract conditions can often appear the same, but, there may be a subtle rewording of clauses which impacts the way the contract is read and administered.

Memorandum of understanding

A Memorandum of Understanding (MOU) is sometimes used as a first stage in establishing a contracting relationship between two or more parties to indicate a common line of action. A MOU can be legally binding providing it:

- is signed by the parties
- clearly identifies the subject matter
- identifies the intention
- there's an offer and an acceptance

In addition the MOU would normally identify a way forward to establishing a formal contract.

A MOU can be useful to clients and contractors when negotiating projects, or trying to resolve disputes, and it serves as an important method to summarise progress made at a meeting.

Letters of intent

Sometimes the client hasn't completed preparing the contract documents but is in a hurry to get the contractor to start work. In this case they may issue a letter of intent, or similar document. These documents are often only a couple of lines long and offer the contractor little protection if things go wrong.

In general I would prefer not to start work without a contract since it's usually difficult to negotiate changes to the contract once the contractor is already on site and spending money.

If a contractor has to work with a letter of intent then the letter should:

- be signed by an authorised representative of the client
- be signed and acknowledged by the contractor
- be dated
- have the correct names of the contracting parties
- refer to the tender conditions
- refer to the contractor's tender submission
- refer to a scope of work
- have the payment terms and conditions
- include a termination clause
- refer to a contract schedule or milestone dates
- include a date by which the contract document will be issued to the contractor

Some Letters of Intent have a set value or are only valid for a specific time period, so it's important to ensure these values or dates aren't exceeded since the contractor may find that they aren't paid for work carried out in excess of that specified.

Payment guarantees

Sometimes the client is required to submit a payment guarantee to the contractor. The contractor should check that:

- the guarantee is received before starting work
- it's made out in the correct names of the contracting parties
- the wording is such that the contractor is able to call in the guarantee if required
- they don't exceed the value specified in the guarantee
- the work doesn't run beyond the dates specified in the guarantee

These guarantees must be kept in a secure place.

Should the contractor have to call upon the guarantee for payment they must ensure that they follow the correct procedures.

Check contract documents

Before signing the contract ensure that:

- the document is the same as the tender document including checking that:
    - no clauses have been altered (unless these alterations have been agreed)
    - the drawings included, or referenced, are the same as those in the tender including the same revision number
    - no new drawings have been added
    - the specifications are the same
- the price is as per the tender and includes all post-tender variations
- the exclusions in the tender submission are included
- the schedule is in accordance with the agreed dates

I've often received documents that had different conditions to those agreed. Even minor alterations to clause wording can drastically alter the meaning and intent. Failure to detect these alterations could result in the contractor committing to undertake work that they've not priced or to conditions they haven't allowed for.

In addition the contract document is often signed weeks or even months after work started so it's important that the document takes cognisance of all delays or variations that occurred up to the date of signing.

Documentation

Correct and accurate documentation is vital on any project, particularly should a dispute arise or when submitting a variation.

Documentation which can become important is:

- photographs showing progress, variations and completed work
- minutes of meetings with clients and subcontractors
- daily diaries
- correspondence
- information requests
- drawing registers
- drawing issue receipts
- site instructions
- notifications
- the contract schedule
- progress updates
- the signed contract document
- the tender (including the submission, correspondence, schedule, drawings, specifications and the post-tender meeting minutes)
- approved day-works sheets

This correspondence should be consistent since I've often found, for instance, project staff enter one thing in the daily diary and then report something else in the weekly report. Claims or variations may fail because of lack of supporting documentation or when there are inconsistencies within the documentation.

Guarantees and warranties

Often contractors incur additional costs when items break-down within the contract guarantee period because:

- the warranty for the part has expired even though it's still within the contractor's guarantee period with the client
- the part hasn't been stored, transported, installed or operated in accordance with the instructions
- someone other than the approved contractor has carried out repairs

Guarantees and warranties must be checked to ensure that:

- they are valid
- apply to the item and provide the required cover
- the guarantee period is sufficient
- the installation, servicing and repair conditions specified in the guarantee aren't violated

Guarantees and warranties should:

- be collected as the project progresses and be kept in a secure place
- have the original handed over to the client with the hand-over documentation, and the client should acknowledge its receipt
- have a copy kept in a file after the project is closed

Disputes

Disputes should be avoided as they:

- are time consuming
- can damage the contractor's reputation
- damage the relationship between contractor and client
- they are costly especially when they become legal
- they may end poorly for the contractor who doesn't receive the full value of their claim

Where possible disputes should be avoided by ensuring:

- there's a legally enforceable contract in place
- the contract offers protection to the contractor
- the contract is well written and doesn't have conflicting clauses or contractual loopholes
- the contractor understands the contract and complies with its provisions
- the contractor communicates with the client and their subcontractors
- the contractor submits and resolves variations as soon as practical
- accurate records are maintained
- there's a willingness to talk and negotiate
- personalities and emotions are kept out of the dispute
- the contractor admits when they're wrong
- the consequences of escalating the dispute are weighed up carefully since the costs of legal action may be more than the outcome is worth
- expert advice is sought when necessary

- the contract is administered in a spirit of honesty and cooperation by all parties

Unfortunately disputes which cannot be resolved do arise and then it's important to follow the dispute resolution process stipulated in the contract. Only as a last resort should you proceed down the legal route. Having said this, though, do not hesitate to ask for a legal opinion or for advice.

Sometimes a dispute is unavoidable, but I have generally found that 99% of variations and claims can be amicably settled without going down the dispute resolution process.

Managers need to be aware of disputes and problems on a project so they can take the necessary action and intervene if required to avoid the problem escalating.

Dispute resolution

There are various options to resolve disputes:

- The cheapest and easiest is to negotiate with the individuals involved with the project, which may require some compromise from the parties.
- If negotiation fails the dispute should be referred to the parties' senior managers. Often the problem is a result of a clash of personality or someone's incorrect interpretation of the contract, and managers can often resolve the issue when these obstacles are removed.
- If the owner of the facility isn't party to the dispute they can sometimes be approached to assist in resolving the dispute.

The contract document often dictates the dispute resolution process to be followed when negotiations fail. This route could include one or more of the following:

- Mediation, where an independent mediator is appointed to bring the parties together to discuss and resolve the problem.
- Arbitration, where an independent arbitrator is appointed to hear evidence and then make a ruling on what they believe is the correct solution. Arbitration is usually not binding.
- Litigation, which tends to take a more legal approach with the process often focussed on the legal rights of a party, without necessarily understanding the project and the impact that the various parties' actions had on the construction processes and schedule.

Arbitration and litigation can be drawn out processes, often taking several years to resolve. They are also costly and may not provide the answer either party was expecting.

Some projects require a dispute resolution board to be appointed at the start of the project which is usually made up of three members (one chosen by the contractor, another by the client, and the third by these two members). Both parties are usually responsible for the costs of the board. The board usually visits the project monthly and they can thus become aware of issues and problems as they arise. Should a dispute arise it's passed to the board for a ruling.

Terminating a contract

Terminating a contract either with a client, a subcontractor or supplier must be done with care. Termination should be used as a last resort since it's usually costly and disruptive to the project.

Before terminating a contract:

- ensure that the other party is in default
- seek legal advice if there is any doubt whether the contract can be terminated
- check that there aren't other alternatives such as providing assistance to the defaulting party or reducing their contract scope
- if possible put alternative solutions in place
- don't take steps which may put your company into default
- carefully consider both the advantages and disadvantages of terminating because the disadvantages often outweigh any benefits of termination
- follow the correct procedures when notifying the defaulting party, including:
    - providing adequate notice in terms of the contract which gives the defaulting party a chance to rectify the problem
    - ensuring the notice is addressed to the correct person, using the correct address, and that it's received by the other party
    - implementing the applicable steps required in the contract

Unfortunately in many instances termination is implemented too late to influence the course of the project.

Joint venture contracts

Before submitting a tender in the name of a joint venture it's important to first seek permission from the client. Some clients may not consider a joint venture if they believe that one of the partners isn't suitable for their project.

A joint venture agreement needs to be drawn up and signed by the parties. This agreement should:

- set out each partner's share in the project
- nominate who will lead the project
- how the project will be managed and administered
- what each partner will contribute in the construction process and how they will be recompensed
- how risk will be shared
- how the joint venture will be funded
- who will supply the project insurances and bonds
- the process for distributing profits or funding losses
- the process for resolving disputes
- the make-up of the executive committee who will run the joint venture

A joint venture is like setting up a separate company and it needs to be done correctly.

Summary

- It's important that there is a contract in place because it protects both the client and the contractor, setting out their rights and obligations.
- Failure to read and understand the contract could lead to expensive and lengthy disputes.
- A contract exists when an offer is made by one party and accepted by another.
- Some clients issue letters of intent to contractors while they are preparing the contract documentation. Contractors need to check these carefully to ensure that they are adequately protected and they should ensure that they don't exceed the provisions in the letter.
- Payment guarantees must be checked to ensure that they are correct and allow the contractor to claim against them should the client not pay for work done.
- Guarantees and warranties should be checked to ensure that they offer the required protection and to check that the contractor complies with their conditions.
- Disputes can, in most cases, be avoided if there is a well-constructed contract in place, which the contractor understands and complies with, and if the contractor communicates with their client and subcontractors, maintaining accurate records and resolving variations as they occur.
- When a dispute arises the contractor must follow the dispute resolution procedures set out in the contract.
- Terminating a contract should be done as a last resort after other courses of action have failed. Care must be taken that all the correct procedures are followed.
- Joint ventures must be set up carefully and there should be a joint venture agreement in place between the joint venture partners.

Chapter 10 – People

Every company depends on people. Even a one person company will outsource some work, with someone preparing their financial accounts and tax, others to assist with legal issues, and, if the intention is to grow, there will eventually be other employees. There's no truer statement than the one which states 'you are only as good as the people that work for you'. I might also add this includes the people that work with you, and, for whom you work.

Construction is a complex and changing process that requires people to change and evolve to suit different projects, clients, locations, challenges and complexities. But to make it more difficult it experiences an inconsistent workforce which varies between regions and cultures. In fact, there are probably few industries that employ people from such a diverse range of cultures, ages, economic means and educational backgrounds, expecting them to work together to successfully deliver a project. To complicate matters further, the workforce is changing, and many cultures and young people have a different work ethic and loyalty from what the norm was a few years ago. Although the processes in construction have remained relatively unchanged for hundreds of years, they are now changing, with new technology, different client requirements, complex regulations and innumerable legislation matters, which often place more emphasis on paperwork and rules than ever before.

In developing and developed countries construction is often viewed as a less attractive vocation than other careers, resulting in a limited pool of skills being available to undertake complex projects. To make matters worse many of the projects are in remote locations and many have unattractive working hours.

The key for any contractor's success is its ability to employ suitable people and retain them, managing them to maximise their worth to the company. Therefore every manager has to understand people, their cultures and backgrounds, and be able to work with them using their strengths, and assisting them with their weaknesses.

Employ the right people

The most important step in having good people is to employ the right people.

- They require knowledge and experience to perform the tasks expected of them. (An experienced building Supervisor is possibly not best suited to supervise the construction of a road.)
- They need to fit in with the culture of the company and must ascribe to the company's values. (It's pointless for the company to set high standards for safety and quality, and then employ a Supervisor who is

unconcerned with these values. They may have all the experience and knowledge for the position, but they will destroy the company's reputation in no time.)

- They should be willing to work in the regions and areas in which the company operates in. (I've seen many personnel unhappy because they've had to relocate their family, or had to work in areas far from where they live. Yet, there are individuals who enjoy working in these regions and others who are willing to relocate their families to remote areas.)
- They must have aspirations which the company can satisfy. (Everyone has different aspirations and not all companies can meet these. Failure to fulfil a person's aspirations eventually results in them becoming unsatisfied and unhappy.)
- Construction is a people business and everyone should be able to communicate and work with others.

Remuneration

Have you heard the saying 'pay peanuts and you get monkeys'? Well believe me it's true, so it's important that the company pays a market related salary, that will attract the right people, and one which is also fair and sustainable.

Most employees look around and compare their packages with what they can get elsewhere, so it's important to do market research to ascertain what other companies are paying. In certain circumstances, with exceptional individuals worth a lot to the company, you may have to even consider paying more than the market value. However, this can later cause problems resulting in these individuals receiving lower increases than the rest of the staff.

Companies don't have an unlimited amount of money to pay inflated remunerations, and excessive salaries can quickly become unsustainable, eventually making the company uncompetitive or unprofitable.

Remuneration should be reviewed at least once a year and increases should depend on:

- the individual's performance
- the average increases in the rest of the industry
- the availability of skills
- if the person's role has changed
- what the company can afford to pay (although care should be taken with this statement because as I've said a company depends on good people and their loss may make the company even less profitable)
- what the company can pay without affecting its competitiveness
- the person's pay relative to others in the company doing the same work
- what the person contributes and their worth to the company
- whether the individual has had a long enough service period to warrant an increase (letters of employment should specify the time within which a review will be undertaken)

I've generally avoided giving one large increase, rather splitting it into a couple of incremental increases.

Pay increases cannot be undone since it's almost impossible to give a person a decrease in salary.

It's also difficult to undo the damage when a person received an increase that was too small since:

- people can become demotivated which could affect their performance
- they start looking for alternative employment opportunities, and once they have decided they are leaving it can be difficult to change their mind

Generally, wherever possible, senior management should personally explain to each employee how their salary has been reviewed, providing the reasons for a low increase or congratulating them on their effort which resulted in them receiving a larger increase. It's often difficult for someone to assess their salary increase by just looking at their payslip, or bank account, because taxes and deductions usually have a dramatic impact and negatively distort the size of the increase.

Sometimes salary packages need to be structured in a way that will incentivise and reward employees, as well as providing tax benefits. Most companies allow employees some flexibility as to how they want their packages structured. A structure that suits a younger person may not be suitable to an older person who could, for instance, want to contribute more to their retirement funding.

Bonuses

Bonuses are an important way to remunerate and retain staff because they can be rewarded in proportion to their efforts and the company's profits.

There are disadvantages to bonuses, one being that when the company doesn't make a profit employees don't receive a bonus, causing discontent. In some cases, the person may have worked hard and had a profitable project while the company as a whole lost money due to other poorly performing projects. In fact, most people expect their bonuses to be the same, or even more, than the previous year's bonus, so a decrease can lead to unhappiness.

Furthermore, when a person calculates their annual remuneration and compares this to packages offered by other companies they seldom factor in the bonus they receive, even if these are a regular occurrence. I worked for many years with a company that paid extraordinarily large bonuses every year. We were fortunate in that the company's profits grew every year for twenty years so bonuses did increase every year. The question to ask is; was the success of the company due to the bonuses? I believe to a certain extent this was the case. The size of the bonuses probably never attracted an individual to work for the company, but it probably helped retain staff, making them feel special and part of the company, striving to make it more successful.

One word of advice on awarding bonuses; it's good practice for a senior manager or the business owner to personally inform each individual of their bonus and to thank them for their effort. This generates far more good will than the person just seeing numbers on their payslip. Having said this, it's just as important when a bonus is reduced, or remains unchanged, that the same manager explains the reasons for this which may include substandard performance.

Share schemes

Share schemes are a good way of enticing employees to stay with the company. Share options or schemes are when an employee is promised shares in the company which they can purchase only in three or five years' time, but at today's price. If the company grows and is profitable the share value should increase during this time, meaning that the employee has a windfall of the difference in the share price at the end of the period compared with the price at the time the share was promised to them. In addition, the individual is paid the dividends the shares earn in this period which can be used to pay for the shares. Of course this assumes that the company's share price has increased in value over the period. If the share price in fact goes down the employee doesn't have to complete the transaction, meaning that they aren't at risk of losing money – they just don't make anything.

A major advantage of the share scheme, aside from the monetary aspect, is that the employee feels part of the company. If they work hard and the company prospers they can usually see a direct correlation in the value of the shares they own. (Unfortunately the share price isn't totally dependent on the hard work of individuals and the success of the company, but it is affected by events in the country and the world which influence the prices of all shares.)

Another advantage is that the person is often less likely to leave as the time draws closer for them to exercise their share options because often the windfall from owning the shares outweighs any monetary advantages of moving to another employer.

Disadvantages are that if the company is very successful and the value of the shares goes up significantly the number of shares that the company can afford to give away reduces. Furthermore the company has only a limited number of shares to work with and when these are gone it's difficult to issue more.

Unfortunately in many countries the tax authorities have clamped down on share schemes and the tax advantages are no longer available.

Share schemes can work well in public companies, but not as well in private companies where there may be restrictions on who can purchase shares and it may also be difficult to calculate the value of the shares.

Other rewards

In this competitive world it's impossible to retain people solely by paying higher salaries than competitors. Firstly, people only interested in money are mercenaries who will happily move from one company to the next for a few extra dollars, leaving you to find a replacement. Secondly, paying the highest salaries isn't always sustainable and could end up making the business unprofitable. Furthermore, there will always be another contractor, desperate for good people, who will willingly pay a premium to entice skilled people for a particular project. It's not to say that salaries aren't important, and you should always ensure that salaries paid are market related – they just don't have to be the highest.

Therefore it's important that people are attracted to the company for reasons other than salary. It's equally important to retain good people. There are different ways of doing this.

Make individuals feel part of the business by regularly communicating with them, discussing the company's future plans and where they'll fit into these plans, thanking them when they have done a job well and making them feel wanted.

Many years ago I remember reading about a tennis tournament whose top prize was a luxury motor car. The tournament attracted more top players than tournaments that offered cash prizes of similar value, or even higher. It's a bit like that in the working environment. Give a person a five thousand dollar bonus and the money goes straight into their bank account where it's used for groceries. The bonus has literally not impacted the person's life. But instead, reward them and their partner with a one week holiday at a luxury resort and they'll probably remember it for many years afterwards. More importantly, their partner suddenly appreciates the company and probably won't be so upset when long hours have to be worked. Unfortunately, in most countries the bonuses and week-long holidays attract tax.

It's often the little things that attract people and it's important to understand what motivates and attracts people since what appeals to a young person freshly out of university is usually not what appeals to a mother with a young family, or to an older person. It's impossible to satisfy everyone, and it's also difficult to meet people's expectations without creating precedents for other employees. However, some items to consider are:

- mobile phones and smart phones
- the use of company vehicles
- personal computers
- flexible work-hours
- additional annual leave
- structuring pay packages to suit the person's needs
- allocated parking spaces at the office
- crèche facilities
- business cards (don't underestimate how important a business card can be because I knew a senior Project Manager who left the company he was working for and one of his reasons was he was never given a business card)
- the layout of offices and desks
- job titles (we don't want to be creating new job titles just to satisfy an individual's ego, but job titles should be similar to other companies in the industry)

Unfortunately construction projects don't always allow for some things, particularly with regards to flexible working hours.

Bear in mind that with younger employees it's not just good enough to provide them with a computer, mobile phone or company car, but it's also important that it's the one that is currently in fashion and with the latest technology.

Company functions are an opportunity for employees to get to know each other outside of the working environment. I remember the first time that the company invited me to an international rugby match and how grateful I was for the opportunity. Of course with all functions it's important not to offend those who aren't invited as there are often a limited number of tickets, or spaces available.

Leave and time-off

I was never happy granting leave but I seldom refused it unless the person had already used more than their leave entitlement. People who take leave frequently are usually more refreshed and happy, and it generally makes their family contented as well.

People should be encouraged to take leave owing to them and they should be discouraged from accumulating it or being paid out for it.

For certain individuals, additional leave may work as a better incentive than a bigger salary.

Motivation

All of the incentives above go some way to motivating employees to do their job, to do it well and to be loyal to the company. However, for a company to be really successful it needs to have employees who are passionate about their jobs and the company, always taking pride in their work.

Part of achieving this is about employing the right people in the first place. Furthermore, the company needs passionate, enthusiastic and knowledgeable managers who will inspire their employees. It's also about keeping individuals motivated and interested in their work which is done by training them, providing new challenges, using them in ways that keep them happy and where they are best suited.

Promoting people

People must be promoted for the right reason.

- Positions shouldn't be created around people or personalities but should be based on what is best for the future of the company. I've worked for companies that have created new positions simply to keep an individual happy and to stop them leaving. These positions weren't necessarily required, and in some instances the individuals weren't even suitable for the new roles.
- People shouldn't be promoted into positions they are unsuited for which often happens when:
    - there isn't a suitable person to fill the post
    - managers try to satisfy an employee's aspirations
    - the company tries to retain an employee who is threatening to resign

  This can be harmful to the individual as well as to the company.

- Promotion shouldn't be delayed because the company fears offending other personnel. If it's right for the individual and the company, then promoting the person should proceed and those that may be negatively affected by the promotion may have to be counselled and have the rationale for the promotion explained.

However, it's often difficult to explain to employees why they haven't been promoted, especially when they see a colleague promoted to become their manager. Sometimes employees have really performed well, deserving promotion, but there are only a limited number of positions available. On the other hand, employees may well be unsuitable, although they probably disagree and have their own expectations. These discussions need to be done sensitively to ensure that the individual is retained within the company and maintains their focus and reliability.

Conversely, there's always the risk that someone isn't promoted because they are just too useful doing what they are currently doing and there's no one suitable to take their place. This is obviously unfair to the individual and probably not good for the company in the long term.

Mentor, train, and develop

Regrettably there's often a shortage of skilled trained people and the only way to have these people is to develop them internally. I worked for a company that had huge success with employing young graduate engineers and growing and developing them so that many eventually became directors of the company. This strategy ensured the tenure of most employees was long, which meant that many of the senior staff had grown up with the company and understood the company systems and procedures, but more importantly, they understood the values and policies of the company and were in fact part of the company's culture.

Training and developing people is one way of retaining employees because they realise that the company is interested in them and that there's a long term future for them. Most people enjoy learning new skills and it stops them from becoming bored.

Of course you also spend time and effort training people who then leave. Don't be offended or disappointed by this, but rather look at it as one more skilled person in the market. If every company trained and developed people there wouldn't be a skills shortage.

Socialise but don't fraternise

I've worked for companies that had a drinks get together on a Friday after work. It's an opportunity for staff to mix with colleagues and personnel from other departments. It's also an opportunity for senior management to meet and talk to staff that they don't usually come in contact with.

However, I've seen managers who take this to the extreme, even becoming inebriated in front of junior employees and losing the respect of their staff.

Common courtesy is appreciated by all staff. I had one business owner who arrived in the office and walked past senior managers working at their desks without bothering to greet them. Apart from just greeting staff it's sometimes good practice to stop at their desk and enquire what they're busy with and if there are any problems.

Family

I found it useful to have an end-of-year function with my staff and their partners. Not everyone attended the functions, but those who did generally enjoyed themselves. In particular, it gave the partners a chance to meet other spouses and to find out that it wasn't just their partner that worked long-hours or away from home. It was also an opportunity for the partner to gain insight into the company and to hear the latest news.

Furthermore, we tried to organise a function or weekend away once a year with senior managers and their partners. We endeavoured to make this a special occasion and something that they wouldn't necessarily have done on their own. These always worked well and were substantially more value to the company than was their direct cost.

Some people may not want to attend these functions so they shouldn't be pressured into doing so.

Celebrate success

Winning a tender usually comes after lots of hard work and probably after losing several other tenders. It's therefore useful to celebrate the awarding of large new projects which provides an opportunity to thank the estimating team and wish the project team success.

When a project is nearing its successful conclusion it's often worthwhile having a function on site where project staff, as well as the Head Office support staff, can be thanked. It's also an opportunity for the Head Office employees to visit the project and gain an insight into the work that the company is undertaking.

Construction isn't an easy business and there are numerous problems encountered on a daily basis, so when there's a substantial success embrace and celebrate it. It helps make everyone feel part of a successful team – success is empowering, infectious and stimulating and everyone wants to be part of a winning team.

Of course celebrations need to be appropriate, possibly limited to some of the more important wins and even restricted to those most affected by the success. Some wins may just call for a congratulatory email to all staff, while others could be suited to a quick drink after work, but a particularly large success may call for a party.

Employing people

Employing new staff can be a costly. These costs include:

- the costs of recruiting such as agency fees and advertising
- time spent on reviewing applications and interviewing applicants
- the new recruit's time in induction and training
- the provision of personal protective equipment, computers and phones

- the wages or salary the person earns when they are unproductive while they learn systems and processes and attend inductions
- medical examinations
- termination should the person prove to be unsuitable or no longer required

It's therefore important that recruitment is done properly and the following should happen:

- ensure that there is a medium to long term need to fill a role
- check that there aren't suitable candidates within the company (including checking with other divisions), or someone that will soon become available
- check whether a current employee can be trained to fill the position
- decide on what the person is required to do and what experience, training and capabilities are required for the position
- decide on the appropriate salary bracket
- advertise the position which could include; asking employees if they know a suitable candidate, advertising in newspapers or asking an employment agency to find suitable candidates
- read through candidates' curricula vitae to assess if they have the required knowledge and experience
- suitable candidates should be interviewed by at least one manager (preferably the person they'll report to) and if possible an industrial relations specialist
- during the interview:
    - the job position and description should be explained
    - the conditions of employment should be discussed
    - assess the candidate's knowledge and experience by asking appropriate questions
    - ascertain the reasons for them applying for the position
    - try and understand their aspirations and gauge whether these fit with what the company can offer
    - review their communication skills
    - understand their salary expectations
    - try and ascertain if their personality will be a good match with the company
    - if necessary ask for additional references
    - ask the candidate what questions they have
    - check the candidate's availability and any future arrangements or plans they have which may impact on their work

In some cases it may be necessary to invite potential employees back for a second interview. It's important that unsuccessful applicants are advised as soon as possible so they can pursue other applications.

Sometimes good applicants may have a number of job offers at the same time. In this case there is always a risk of a bidding war erupting between the different companies with each offering a larger salary than the other. In these circumstances it's always essential to ensure the salary offered is sustainable for the company and is market related. As discussed earlier it's also important to stress to the applicant the other benefits of working for the company. If a candidate joins another company only on the basis of a higher salary then maybe they were the wrong person for your company.

Probation period

It's important that all new employees work a probation period. The employee's performance must be reviewed before the end of this period to ensure that their performance matches the company's expectations. If it doesn't the person should be called to a meeting and informed why their performance is unsatisfactory and either their employment should be terminated or their probation period extended, providing them with a chance to improve. This meeting should be documented. At the end of the revised probation period the person's performance should be reviewed, and if it is still unsatisfactory their employment should be terminated.

Unfortunately in many cases the probation process isn't handled properly resulting in poor quality people remaining in the company's employ, causing additional expense when their employment is eventually terminated.

If the person was engaged through a recruitment agency and their employment is terminated during their probation period it's usually possible to get the agency to supply another candidate without charge or to refund their placement fee.

Bursaries and in-service training

Providing bursaries to students at colleges and universities is a useful way to recruit young qualified people. It's important that candidates are carefully selected to ensure that they not only succeed with their studies but will eventually be suitable and a good fit for the company. There will always be some successes as well as failures in the process, so be prepared to give more bursaries than the positions you require filling.

Many college and university courses require their students to spend time with a company in their chosen field in order for them to obtain practical experience. I've witnessed some students unable to complete their courses because they couldn't find a company to give them the required experience. This isn't a good reflection on the industry and is unfair on the students.

I've often had good results with these students since:

- they are relatively cheap to employ
- it gives the company an opportunity to understand the capabilities and personalities of the students, and if they are suitable to offer them long term employment once they qualify
- many were useful on the project
- it gives the students an opportunity to see whether they would fit in with the company if they were offered long-term employment

- many students appreciate the opportunity of employment and go on to become loyal employees

It's however important that students aren't just given boring mundane work. They need to be given an opportunity to learn things and staff must be encouraged to support them and to pass on their knowledge.

Sometimes it's difficult to get students onto projects for a short duration, particularly projects in remote regions or which have lengthy processes to gain access.

Contract staff

Sometimes it's possible to employ staff on contract for short periods to assist with short term shortages.

Advantages of employing contract staff are:

- they can be employed for a limited period to deal with specific problems
- they are normally easy to replace if they are unsuitable
- they usually don't require notice when their services are no longer required
- it's possible to secure skills which the contractor wouldn't normally employ

There are disadvantages:

- they are usually more expensive
- they are often employed through agencies which add their overheads and profits to the person's rate
- they don't have an allegiance to the company resulting in limited loyalty
- many don't ascribe to the company's values
- they are often paid by the hour, so may be happy to drag a task out as long as possible to maximise their pay
- sometimes management forgets the person is employed on contract and the worker ends up being employed for several years at great expense
- care must be taken to check that the hours worked are the same as those that are invoiced

Contract staff do have a place, but I wouldn't recommend a company uses them on a regular long-term basis. Where possible I would always try and ensure that permanent employees were placed in the important roles, and in particular in positions where they are dealing directly with clients.

Employing workers permanently or for a limited duration

Workers can be employed on a permanent basis or for a limited duration for a particular task or project. In some countries it's even possible to engage a labour broker who will supply the workers for as long as they're required.

Advantages of employing workers on contract for a particular project are:

- when their specific skill is no longer required their employment can be terminated

- when moving to a new area or region it's often possible to employ people locally which saves transporting and accommodating workers from other localities
- sometimes when permanent workers are transferred to a different region they expect to be paid additional allowances
- wages often vary from region to region and local workers may earn less than their counterparts from other regions
- the contractor doesn't have to keep a large pool of resources employed

Disadvantages of employing workers on contract are:

- they may have little loyalty to the company
- they usually aren't familiar with the company's processes and procedures
- they will be untested and it could take several weeks to assess their level of skill
- they won't be familiar with the company's safety and quality standards
- the company will spend time and effort on training people who are employed for only a limited duration
- the company may be unable to find skilled people when they're required

Probably the best solution is to employ a few skilled tradespeople on a permanent basis and transfer these between projects so they form a core for the team on the new project. The balance of workers can then be employed locally or from other sources.

Industrial relations policies

Industrial relations policies set out the employer's and employees' obligations. The company should have a clear industrial relations policy which would include some or all of the following:

- inductions
- work hours
- safety procedures
- disciplinary procedures
- grievance procedures
- anti-discrimination policy

It's important that these policies are understood by all employees and that management and supervisors apply these in an even and fair manner.

Many industrial relations problems occur in companies due to policies not being applied correctly and in some cases by even being ignored.

In some countries and states it's mandatory to have these policies in place.

Employment contracts

All employees should have an employment contract. The contract should include:

- their expected duties or job description
- the place of work
- the date of commencement

- their remuneration
- applicable allowances
- their hours of work
- notice period for termination for both parties
- reasons for termination
- the duration of employment
- leave entitlements
- probation period and the notice period for termination in this period
- general conditions of employment
- workplace rules
- use of company equipment and property
- definitions

Care must be taken to ensure conditions and remunerations for like grades of workers are similar. In addition, the conditions must comply with the legislation and existing labour and union agreements.

From time to time personnel may be transferred to projects with different rules and conditions. In these instances employees should be given a secondment contract with these revised conditions. However, the secondment contract shouldn't have conditions inferior to the individual's basic conditions of employment. On completion of the project, employees should either be given another secondment contract for a new project, or issued a letter stating that their employment conditions have reverted to those of their original employment contract.

Employees must sign all employment and secondment contracts and they should keep a copy, and a copy should be kept by the company

Shop floor and union agreements

Different regions and countries have rules governing basic employment conditions and the company must ensure that these are applied and abided by. In addition, workers could belong to a union and it's then usually necessary for the company to have an agreement with the union which governs the workers' conditions of employment. When dealing with a union it's important to ascertain that the union is registered and that it represents the employees. Sometimes, to complicate matters, employees belong to different unions and the company has to engage with more than one union.

Once a union has sufficient representation (usually 50% of employees) they will approach the company to negotiate employment conditions. Some of these can add significant costs to the company and make it uncompetitive, particularly when tendering in other regions or against contractors who don't have union agreements. The contractor must negotiate terms which are fair and equitable to their workers, maintaining worker harmony, without impacting on the company's competitiveness or profitability.

Where possible the company should try and structure these agreements for a minimum two year term, but preferably for a longer period. These negotiations usually take a huge amount of management time and nobody wants to go through the process every year. Furthermore, long term agreements give the company certainty as to what rates should be used in their tenders.

If employees aren't represented by a union it's sometimes possible and advantageous to sign a shop-floor agreement with the employees. Workers elect suitable employee representatives to negotiate on their behalf. When the negotiations are complete it's advisable to get an industrial relations lawyer to draft a legally binding agreement which all parties can sign. In many cases this document has to be registered with the relevant government agency.

In all cases when negotiating with unions or employee representatives it's advisable to seek expert assistance and guidance. Badly negotiated or worded agreements can turn out to be costly for the company.

Different cultures and ethnicities

People working in the construction industry usually come from diverse cultures, backgrounds and ethnicities with different work ethics and ways of doing and saying things. Managers need to be aware of these differences and take them into account when communicating and directing people. Often minor misunderstandings created by these differences can escalate into major industrial relations incidents, disharmony and loss in productivity.

By understanding the different cultures it's sometimes possible to improve people's work performance and productivity by adapting the way you communicate and deal with them to suit their backgrounds. What is important to someone from one culture may not be as important to someone from another culture. What is amusing to one person may be offensive to another from a different nation.

Discrimination

It's important not to discriminate and to be aware that many people can be sensitive and easily take offence to something that they perceive is discriminatory. Sometimes, they may even deliberately take something out of context and twist it so it appears to be discriminatory and they can create an incident. Discrimination can lead to expensive legal disputes or work stoppages.

Discrimination can take many forms such as; gender, racial or personal.

The company should have clear anti-discriminatory and sexual harassment policies and action should be taken when discrimination or sexual harassment is witnessed or reported.

Indigenous and local people

In many countries and regions the employment of indigenous or local people is actively encouraged and in some cases it's even legislated. Management needs to ensure that the employment of indigenous and local people is actively encouraged and furthermore that these employees are retained for as long as possible on a project provided that they are suitable for the positions occupied. Once recruited indigenous people should be managed by staff who are sensitive to the need to retain them and who understand how to manage, train and develop them so that they can become successful employees.

Delegate

I worked for a business owner who kept taking on more and more responsibility with the result that many of the tasks weren't completed which frustrated staff as well as clients.

It's important to be able to delegate certain tasks and actions to others. However in doing this ensure:

- that the person is capable of doing the task
- the person is appropriate for the task (you don't want to delegate a junior engineer to attend an important client meeting on their own)
- they understand what's required
- they have all the information and resources to do the task
- if they don't have the experience or knowledge then spend time to teach them, or ask someone else to show them
- follow up to ensure that the task is done correctly

People will not grow and develop if they are not shown how to do a job and if they aren't delegated more and bigger responsibilities.

Supervise and Manage

Many businesses fail because people aren't supervised and managed correctly resulting in:

- personnel performing poorly
- people not doing tasks they are meant to, or doing them poorly
- tasks being left undone because they aren't allocated to a person
- staff not working as a team with a common goal
- personnel becoming disillusioned and unhappy at the lack of guidance and order

It's therefore important not only that managers manage the people reporting to them but that they also ensure these people manage the teams reporting to them.

Discipline and commitment

Poor worker discipline, performance or lack of commitment shouldn't be tolerated since this leads to poor productivity. Discipline must be applied consistently from the start of every project and uniformly across all projects. Problems often result when workers are transferred from one project where the company rules have not been applied correctly to one where the Project Manager enforces the rules. Employees don't understand why the rules differ from their previous project.

It's important that disciplinary procedures are carried out correctly:

- the process must be recorded in writing, even when verbal warnings are issued

- employees must be given sufficient notification of formal disciplinary hearings
- care must be taken to explain the problem or transgression and if necessary an interpreter may have to be provided
- the employee may bring counsel to a meeting depending on the worker agreements and legislation
- the employee normally has a right to appeal the findings of the hearing
- records of the process and notifications must be kept in the employee's record folder and copies kept at Head Office
- discipline must be implemented in an unbiased manner

Regrettably, some companies aren't consistent in their application of discipline allowing some staff and key personnel to escape censure, making it difficult to apply discipline properly to other personnel.

Poor worker performance

Often workers perform poorly and it's important to understand the reasons, which may be due to:

- the person lacking experience or knowledge
- a lack of understanding of the instructions
- poor supervision
- personnel not being utilised correctly
- poor morale
- personnel not understanding what's expected from them
- the employee being unwilling to do their work properly

Once the problem has been identified the employee should be:

- counselled
- redeployed if necessary
- if their performance doesn't improve the disciplinary process should be followed and if necessary the person's employment terminated

Many companies transfer substandard employees from one project to another without taking appropriate action to improve their performance. This causes poor productivity and results in frustration amongst the supervisory staff and other workers who have to work with them. The longer the employee remains with the company the more difficult it becomes to terminate their employment.

Lead by example

Some company owners and managers set rules which they themselves don't follow. I've always been conscious of how observant employees actually are. They always seem to know when managers take an extended lunch break, arrive late or come to work intoxicated. If they perceive this to be acceptable behaviour they see no reason why they shouldn't do the same, and when disciplined for this behaviour they are quick to remind the managers that they committed similar offences which weren't punished. It then becomes difficult to apply discipline properly because employees will say they are being discriminated against – with one set of rules for management and another for workers.

Feedback

Employees should be given feedback on what management consider their weaknesses are, or areas that require improving, otherwise most won't know where they're going wrong. In fact, many will probably believe they're doing a good job.

Positive feedback is just as important as negative feedback.

Negative feedback must be given in a constructive manner; highlighting steps which the employee needs to implement to improve their performance.

It's good practice to have formal annual or bi-annual performance reviews which are documented.

Know and understand the team

I've often been successful because I was able to work with staff I'd worked with for several years so that I understood their strengths and weaknesses. In fact the few projects that weren't a success were often because they had new staff whose abilities I didn't fully understand.

By understanding their strengths and weaknesses I was able to:

- use and work with their strengths
- allocate the staff with the right capabilities, knowledge and experience to a project
- support their weaknesses
- put balanced teams on projects that supported each other
- mentor and coach them that they eventually overcame their weaknesses
- promote individuals to new positions knowing that they were capable of filling the role

To properly understand a person's ability it's important that they remain with the company for an extended period, and, that managers are able to spend quality time with them, to mentor and coach them.

Hiring and firing

Some companies are continually hiring people and then dismissing them because:

- their planning is poor and they're employing too many people or people with the wrong skills
- they're employing unsuitable people
- they're getting rid of people without checking if they're required elsewhere in the company
- they don't bother to counsel or train workers, rather dismissing them as soon as it appears that they may be unsuitable
- they go from a situation of too much work to one of too little work

The problem with the mentality of hiring and firing is:

- as discussed earlier it's expensive to employ people
- there's no continuity or loyalty in the workforce

- workers have poor morale and are never sure if they will be the next to be dismissed
- the company develops a reputation of not employing people for the longer term

Labour disputes

It's best to solve labour problems before they result in a dispute. Even the smallest problem left unresolved can quickly snowball into a full labour dispute affecting the entire workforce on a project. This however doesn't mean that you should give in to unreasonable demands from the workforce. Rather, care should be taken to treat the workforce fairly within the structure of the workplace and project agreements. To give into requests additional to their entitlements often results in the workforce coming up with further demands.

Unfortunately most companies will face a labour dispute at some stage that degenerates into strike action. It's important that management is aware of the procedures to follow should there be a strike, which may include:

- notifying company senior management and the human resources department
- notifying the client
- ensuring no employees are in danger
- checking that property and equipment are safe
- safely shutting down systems and operations
- meeting with worker representatives to understand the problems
- resolving issues if they are simple
- encouraging workers to return to work while their grievances are attended to
- ensuring that procedures are followed (these may depend on whether the strike is legal or not as well as on the labour agreements and legislation)
- if there's a risk to property or life, notifying the police
- ensuring staff remain calm
- avoiding inflammatory statements or antagonising the workers
- if workers are unionised, make contact with union leaders to request assistance with resolving the issues
- ensuring only authorised representatives from the company deal with the press
- notifying management, the client and police as soon as workers return to work
- resolving any outstanding grievances promptly and keeping workers' representatives informed of the progress

Summary

- Employing the right people is vital to a company's success.
- Good people need to be retained which is done by:

    - paying them a market related salary
    - incentivising them with bonuses and share schemes
    - making them feel part of the company
    - finding other rewards that are important to the individual
    - providing sufficient leave
- Employees should be motivated and passionate about their work and the company.
- Employees must be promoted for the right reasons.
- Mentoring and training motivates employees and it enables the company to develop a more skilled workforce.
- Inviting employees' partners to company events can be useful.
- Celebrating success is not just an opportunity to thank those who created the success but also makes other employees feel part of a winning team.
- The employment process needs to be done with care ensuring that the potential new recruit has the right attributes and that they fully understand the nature of the company and what role they will be filling.
- Probation periods must be managed correctly ensuring new employees are suitable; if they aren't they should be counselled so that they either improve or are released from the company's employ.
- Offering bursaries and providing in-service training to students is a useful way of attracting new employees and it also contributes to the industry.
- Usually companies have an option to employ staff and workers permanently or for a limited duration only. There are advantages and disadvantages to both options.
- All companies should have industrial relations policies and procedures in place which are available and understood by all employees.
- All employees should have a valid employment contract which may be amended from time to time when employees are seconded to projects with better conditions to those in their contract.
- Companies may have to agree working conditions with unions or employee representatives.
- Managers should understand the cultures of the people working for them, and deal with them in an appropriate way.
- Discrimination of any sort shouldn't be tolerated.
- The employment of indigenous and local people should be encouraged.
- Managers must be able to delegate.
- Employees must be supervised and managed properly so that they perform their duties correctly.
- Discipline must be enforced fairly, appropriately, consistently and in accordance with the company's policies across all facets of the business.
- Poor worker performance shouldn't be tolerated and it's important that steps are taken to rectify it.
- Managers must lead by example.
- Employees must be given regular feedback about their performance.

- By understanding employees' strengths and weaknesses it's often possible to place them in positions best suited to their capabilities and provide support to those that require it.
- Companies should avoid hiring and dismissing people on a regular basis.
- Labour disputes must be managed appropriately and with care.

Chapter 11 - Managing the Company

Many owners and managers of construction companies have sound contracting knowledge and experience but have little experience in running a company. Managing a company requires knowledge of financial, legal and tax systems, an understanding of environmental, safety and labour legislation, compliance with building codes, registrations and planning permissions as well as people skills, being able to employ suitable staff, managing and directing them, as well as being able to deal with clients, designers, bankers, suppliers, subcontractors and members of the public.

Generally there are a myriad of rules and regulations companies must comply with. I'm not going to dwell on the legal and tax requirements for running a company as these vary from state to state, between countries, and regularly change, so it's best to consult experts.

Requirements to manage

Managing a construction company or division isn't easy. Individuals need to:

- have initiative
- be able to deal with different types of people, with different education levels and from different cultures and ethnicities
- be able to make tough decisions
- be passionate about what they do and about the company
- be collaborative and able to share ideas
- be able to communicate
- have the respect of their employees
- listen to employees and their suggestions
- have a good level of detail
- absorb stress
- be adaptive
- be able to delegate
- work within a team
- be able to deal with many problems and tasks almost simultaneously
- have a situational awareness capable of seeing potential problems
- have a sound technical knowledge of the industry
- have an understanding of the market place in which they are operating
- understand the company systems, procedures and policies ensuring that they are followed
- have a vision for the future of the company

To manage a construction business means you have to expect the unexpected and be able to rapidly institute a contingency plan.

Open door policy

Managers should have an open door policy so that employees are free to discuss problems with them. Clearly this has to be done within reason and employees must realise they cannot interrupt meetings and phone calls and that sometimes managers need some quiet time to attend to problems and to answer correspondence. It also doesn't mean that employees should bypass their direct managers and take their problems to a more senior manager.

Part of the open door policy does extend to ensuring that the office is suitable for one-on-one meetings, or even meetings with two or three individuals.

Case study:

I knew a business owner who had a small office with a desk and two chairs. Any meeting in his office ended in an awkward situation with him and me sitting facing each other with our knees touching. Needless to say if there were three of us it became even more crowded with three pairs of knees touching. This really wasn't conducive to having a meeting, which I think was probably the idea in the first place.

Furthermore it's preferable if senior managers have an office with a door so that confidential discussions with clients or employees aren't overheard. I know many people are in favour of open plan offices and would disagree with me.

Reports

I've seen many managers demand extensive and lengthy reports. One company I worked for had a monthly report in excess of a hundred pages long. Was this efficient? Definitely not! It took a number of people several days to put together, time that could have been better spent earning money on the project. When the report becomes so lengthy, and takes so much time and effort to put together, the quality of the information in the report is often questionable since in the rush Project Managers insert any information, and even just make it up, so they can complete the report. Of course the question is, who actually reads the contents, and more importantly who deals with items which should be actioned? In most cases a substantial portion isn't read and the time and effort spent on compiling the report is wasted.

I was fortunate to work for a company that produced a simple three page monthly financial report which was just as effective.

Meetings

Meetings can be a big time waster. Some companies have weekly or fortnightly meetings. Many of these can be several hours long. Some meetings are worthless when important participants aren't present.

- Meetings should be set for a fixed date and time as far in advance as possible.

- They shouldn't be rescheduled unless there's absolutely no alternative. I've frequently had managers reschedule meetings at the last minute, inconveniencing other participants who then have to reschedule their calendars at short notice. Often some attendees are unavailable for the new time. I found that once a regular meeting is rescheduled it's easy for it to become a habit to keep rescheduling it, and eventually the meeting loses its importance.
- Meetings should run to an agenda.
- Meetings should be minuted and the minutes should be circulated within a few days of the meeting.
- The minutes should allocate responsibilities for actioning items and a time for them to happen.
- The chairperson needs to exercise control to keep discussions to the point, focussed and relevant to the topic.
- Meetings should be scheduled for a time that's suitable for most participants taking into account the location of the participants and their current projects.
- Participants should arrive on time. I cannot count how many hours I've wasted waiting for meetings to start because they've been delayed by tardy participants.
- If someone is unable to attend they should excuse themselves beforehand so the meeting isn't held up waiting for them.
- Attendees should be restricted to those who can contribute to the meeting or who will benefit from attending.
- It may be necessary to review attendees, the agenda and the schedule from time to time to maximise the benefit of the meeting.

I've found that many items discussed in meetings could have been discussed on a one-on-one basis with the particular individual, thus avoiding wasting time in the meeting.

Attending to tasks

Managers and business owners usually have a busy schedule and it's sometimes difficult to get to everything requiring their attention. Unfortunately some tasks have to be done, and failure to do them could result in payments being missed, tenders not submitted on time, or clients being upset by non-attendance to their problems. Where possible, to ease these pressures, functions and duties may have to be delegated to suitable staff. In addition, managers may have to schedule their activities around the important tasks that have to be done on a regular basis.

Learn to say no

Managing a company successfully means you have to learn to say no. Obviously this should be done politely and if possible with a reason. As discussed, certain clients can be quite insistent that you should undertake their projects, some of which might not be viable, too risky for the company, or, the company may not have suitable resources available. Clients will be unhappy when the contractor declines their project, but they'll be more upset if the project is executed poorly.

Employees can also be demanding and it's often easier to say yes when someone wants to take time off or asks for a favour or special treatment. Giving in to these demands usually isn't a problem in a smaller company, but as the company grows these demands can create a multiplying effect with other employees, who will all want similar treatment.

The art of persuasion, negotiation & communication

Managers have to communicate, persuade and negotiate on a daily basis. The way they do this will often depend on who they are dealing with and what the situation is. Management will normally deal with workers, staff, subcontractors, suppliers, the client, the client's team (including designers, project managers and architects), local authorities and members of the public.

Communication should:

- be civil
- be clear and concise
- be persuasive and forceful enough to ensure that instructions are followed
- achieve the best outcomes for the company
- be effective
- take into account relationships
- take into account the level of understanding the other person has
- not be condescending

Good communication is vital to the success of the company. There are courses which can improve the level of written and oral communication, and consideration should be given to senior personnel attending these courses.

Communication is not just about giving and receiving instructions, it's also about keeping the various stake holders such as staff, workers, subcontractors, suppliers, client representatives, the client and neighbours informed. The amount and level of communication will vary according to circumstances and to the level of the individuals. Communication is often best given verbally, which could be at meetings, directly one-on-one, or via telephone. Sometimes the communication is in the form of letters or memos that may be addressed to specific individuals, or may be in the form of generalised memos handed to all relevant parties, or displayed on project notice boards.

When issuing letters, emails, memos and verbal communications, thought must go into the type of communication. Inappropriate communications can cause irreparable harm to the company, individuals, and to personal reputations, and often things are said or written in haste, which are regretted long after the event.

All staff must be aware of the communication protocol, and correspondence must be carried out in a professional manner. All correspondence of a contractual nature, or correspondence that involves a financial matter, should be reviewed by the relevant managers.

Stand up for your team

Many managers don't stand up for their staff, often allowing the client to bully them, sometimes even allowing clients to remove staff from a project, when in many instances the client is wrong and the person was simply protecting the contractor's rights.

I've generally tried to stand up for my staff when they were in the right, and have on occasion persuaded the client to retain the person on the project. Of course, always investigate the issue to ensure that your employee is right, since I have on occasion been lied to by employees, and, if I had supported them I would have been wrong.

Decision making

Managers have to make decisions on a daily basis that affect the running of the company as well as individual projects and tenders. Many of these decisions can have a significant impact on the finances of the company, in some cases the livelihood of people, and even the lives of people.

It's important that major decisions:

- are thought through
- consider all the available information
- are made for the correct reason
- are made timeously

A decision made late or not at all can often be more harmful than a bad decision. In construction there's no place for procrastination since this can be costly. Clients and staff expect managers to make timely but sound decisions.

Organising and analysing information

Managers often have a mountain of information to analyse from the many different projects they're responsible for as well as their other duties. It can be difficult to sort out which are the most important issues, then analyse their accuracy, interpret them and act appropriately. The fact that a project is in serious trouble is often seen far too late by many managers.

Reporting systems need to be simple and easily read, with salient points readily visible. Of course nothing beats having regular one-on-one meetings with staff or visiting projects, ensuring that pertinent questions are asked.

By appropriately delegating certain responsibilities it's possible to reduce the volume of information that needs to be reviewed and assessed.

It's also important that subordinates understand what information needs to be passed on to their manager and for them to know that when in doubt their manager will always be willing to assist and advise them and answer any questions or concerns they may have.

Organisational structure and reporting lines

All construction companies, other than micro companies with only a couple of employees, should have a formal structure so that staff-members understand their responsibilities, their limits of authority, who reports to them and who they report to.

This structure will change and must be adapted as the company grows.

The structure doesn't have to be distributed to everyone within the company, although in a large company it's useful for everyone to be aware of the different departments and divisions in the company, knowing who they should contact for information or when they have a problem.

Company overheads

I've discussed making projects more profitable by putting in controls and managing resources, but it's equally important that the corporate office is also run efficiently. Many construction companies struggle financially because of the high operating costs of their Head Office. Often when companies grow they move into a newer, more lavish office block, in a prestigious location. Unfortunately the rents for these types of accommodation are usually high. Don't get me wrong, it is important to work from offices which give a good impression to clients and are comfortable for personnel to work in – but they don't have to be opulent or extravagant in terms of their facilities or location.

When choosing offices consideration should be given to future growth. This doesn't mean you should rent or buy large office space for a handful of employees, but it's useful to have a few spare offices which can be used when the company hires new employees. Companies with insufficient space end up having to relocate every few years or have to operate from several different locations which can be disruptive and costly.

Too many staff in the Head Office creates large additional costs. Small and medium sized companies may have to employ people who are willing to multi-task, since it's not possible to employ individuals for each task.

It's also important to have sufficient controls in place to limit unnecessary or wasteful expenses. These controls may include power usage, printing and stationery, refreshments, telephone usage, company credit cards and travel expenses. Of course it's important that the controls don't become overly restrictive, impacting productivity.

As with the projects, all expenses should be scrutinised and approved by senior managers. The Head Office must have similar controls to the projects to ensure staff are productive, they're managed properly to their full potential, are adequately trained for their tasks, they submit appropriate leave forms when required and that their time sheets are correctly completed.

Departments

When a company starts off the owner is often the manager, the financial officer, the clerk, bookkeeper, Project Manager, Supervisor and Estimator. As the company grows and expands, the owner can no longer do everything and employs others to do these tasks. As the company grows further, there isn't just one bookkeeper, clerk or Estimator but a number of people employed to do these functions. The owner can no longer manage all the different functions within the company and has to employ managers to do this. These managers report to the owner, chief executive officer or managing director.

Eventually the company becomes organised into different departments with different responsibilities and managers. As the company grows further it may become necessary to reorganise some of these departments and some of their duties may be moved to other or newly created departments.

All of these departments need to function together as one unit to make the company successful. Each department has to support the other departments and the projects.

Departments should never be created just to satisfy someone's ego or to replicate structures in other companies, but should be put in place only when needed and when it will create better service within the organisation.

Companies need to guard against those who want to build large departments to emphasise their importance. Each department should prepare an annual expenditure budget which they must justify, and then adhere to during the financial year.

Divisions

Also as companies grow they may form different divisions. These divisions are usually organised around a particular function or trade. Many construction companies have building, roads and civil divisions. Sometimes divisions are focussed on a particular region or country.

The purpose of creating divisions is so that:

- when the company grows, and one manager can't look after all the projects, the company is split into different divisions and the workload is shares between a number of managers
- when staff have knowledge or expertise in a specific field of construction it's appropriate that they are employed by a division that can effectively use these attributes (for instance a building Supervisor is best suited to doing building projects)
- clients know that they are dealing with a company or division that can deliver the type of work that their project demands
- where employment conditions and rates of pay differ between areas and types of construction operations it is possible to separate employees into various divisions where they can have different employment contracts from others in the company
- where types of construction have different rules to others, the project is allocated to the division that has experience of these working conditions (for instance constructing a building in an oil and gas plant is very different to constructing one in the city, and those unfamiliar with the rules in an oil and gas facility could end up pricing the tender incorrectly and will probably make numerous mistakes during construction)

Case study:

Our company's civil division constructed many projects for the mining and resources industry. However, once when the division was awarded the construction of a new coal process facility they didn't have the resources to undertake the project, so the company's building division elected to take on the project. Unfortunately they soon experienced problems. They weren't used to the rigorous safety requirements or stringent quality documentation demanded on this type of project. They were also surprised by the length of time taken to get resources through the induction processes and on to site.

In addition their personnel weren't happy to live away from home, which most civil projects demand.

Staff rapidly became disillusioned and morale was poor which adversely affected productivity.

Management made excuses for the project saying it was a particularly difficult client. In truth, they were no more difficult than most clients in that industry, just that the building division staff weren't used to working under these conditions.

However, by the same token, if the civil division was asked to build an apartment block in the city they probably wouldn't have been able to perform as well as the building division could as they weren't experienced in that kind of work.

Sharing between divisions

I was fortunate to work for a successful contractor that had more than a dozen divisions. Some of the success was a result of the divisions working together. In many companies the various divisions operate in their own 'silos' and are so compartmentalised that they seldom even talk to each other. In fact, I found it so bad in some companies that the divisions subcontracted portions of work out to other companies rather than asking another division within the company to undertake the work. Often these same companies had one division hiring personnel while another division was terminating staff.

Wherever possible divisions need to share resources and manage projects so that as much of the work as possible is done within the company.

It shouldn't be about personal egos or how much money individual projects or divisions make, but rather about the success of the company as a whole.

We often formed internal joint ventures between different divisions which enabled each division to maintain independence, but at the same time to share in the profits and reputation created by the project. We also seconded resources to other divisions when we didn't have work for them.

Managing safety

It's important that senior management are safety conscious, enforcing the safety rules, setting a good example and mindful not to allow unsafe acts to happen in their presence. The company should have a clear well-defined safety policy and a set of procedures which should be adhered to and enforced on all projects even when they are more stringent than the client's requirements.

Safety should be the first item for discussion at every management meeting and the safety statistics for all projects and those for the whole company should be discussed.

Good safety should be rewarded while poor safety should be corrected immediately. Safety is about changing the mindset of all employees so that they think and act safety through all of their actions at work, while travelling to and from work, as well as at home.

A culture of reporting and investigating accidents and incidents must be encouraged so that lessons can be learned to prevent further similar incidents. It's important that senior management are made aware of serious incidents and accidents as soon as they occur. There's nothing worse than a client phoning the contractor's senior management to complain about a serious incident or accident on their project of which management were unaware of because the Project Manager hadn't reported it.

Tender systems

I've discussed tendering at length in a previous chapter, however, I would add that when doing tenders, other than basic ones, consideration should be given to using a propriety tender system. Although these systems can be expensive to purchase, or hire, they not only save time and improve the accuracy of the tender process but they also provide other useful information, some of which can be included in the tender submission and some that can assist the Project Manager in running the project.

Documentation

Electronic documentation is a wonderful asset because it can usually be rapidly retrieved and be accessed from remote locations. However, it's important that electronic documentation is stored in a central location and is backed up to a second secure remote location. I've had occasion when managers have had their computers stolen, or their computers have malfunctioned, and they've lost valuable information which has disrupted their work and taken many hours to recreate.

Electronic document control systems should:

- be simple to use
- compatible with existing software
- have sufficient storage capacity
- be password protected to protect sensitive information
- be retrievable
- be distributable

Insurances

All companies should have adequate insurance cover which should include:

- worker's compensation
- third party liability
- cover for vehicles, plant and equipment (check that external hired equipment is also covered where necessary)
- insurance of the works
- if there's design involved, professional indemnity insurance

- transit insurance for when expensive items and equipment are transported

Failure to have adequate insurance in place can be costly. Furthermore, ensure that the premiums are paid on time since failure to do so will invalidate the insurance.

Most insurance policies are renewed annually and it's prudent to negotiate revised premiums, but in doing so carefully comparing quotes from different insurers to ensure that they provide adequate cover and that their 'small print' will not invalidate any of the claims. Depending on the risks it may be possible to negotiate lower premiums by accepting a larger excess or deductable amount.

It's important to regularly review the company's risks and exposures and to check that they are adequately covered by the insurances that are in place. The company must be neither under insured nor over insured. (Being underinsured may mean that items aren't covered, or are only partly covered, should there be a loss event, while being over insured will mean the company pays unnecessary insurance premiums.) Therefore, update insurance policies regularly to take into account changed circumstances, growth in the company and additions to the equipment fleet.

When starting a new project consult your insurance broker to confirm that you are adequately covered under the existing policies. Additional insurance may have to be bought to cover particular risks which are excluded from existing policies, such as fire or flood.

Check what the client's policies cover, since you don't want to buy additional cover for an event which is already covered by the client. There may also be additional insurances required by the client, or they may require insurance cover for a higher value than the company's existing policies. In addition there may be statutory insurances required which might vary between projects and regions.

Permits, licenses and registrations

In order to operate in most regions, states and countries, companies may require a number of licenses, permits, and registrations. Where necessary obtain legal advice when entering a new region to ensure that the company has everything in place.

Maintain a log of all registrations to track and ensure that they are all valid. Note that many of the permits and registrations have to be renewed annually.

In order to maintain some of these permit conditions, it may be required that certain employees have a particular license or qualification. It is thus important that these qualifications are kept current and that when staff resign there are still sufficient qualified people to comply with the registration requirements.

Have copies of the registrations readily available. In foreign countries on remote sites it may be necessary to hold copies of the registrations on the project site in the event that the local authorities carry out an inspection.

Failure to have these in place may result in the contractor being fined and the project stopped, which could result in serious contractual consequences and costs.

Guarantees

In general clients require a bond or surety to be in place for the duration of the project, including the maintenance period which is usually at least one year after the work has been completed. Most companies have limited facilities to obtain bonds and guarantees because these depend on the amount of security the company can provide. In many cases banks and insurance companies require the security to be a cash deposit, which ties up money, and impacts the company's cash flow. When the company's guarantee facilities are exhausted they won't be able to obtain further guarantees, resulting in the company being unable to undertake further projects.

In addition, guarantees cost money. The amount depends on the value of the guarantee as well as the length of time for which it's required. When guarantees aren't returned on time additional fees are charged.

It's vital that companies track all guarantees, ensuring that they are returned by clients as soon as practical so they can be given back to the bank, freeing up the facility for new guarantees. To facilitate the return of guarantees companies need to close out projects as soon as possible. Sometimes minor items aren't completed, or the final documentation isn't submitted, which prevents the client from issuing the completion certificate, thus delaying the release of the guarantee.

At the end of the maintenance period it's equally important to request the client to prepare a final inspection defects list which the Project Manager needs to attend to as quickly as possible. This period often drags on needlessly because Project Managers forget to follow up with the client, or they take longer than necessary to complete the defects. In some cases Project Managers even forget to request the return of the guarantee.

Policies and procedures

All companies should have policies and procedures in place to cover amongst other issues:

- industrial relations
- safety management
- quality management
- environmental management
- the use of company equipment
- financial controls
- sexual harassment

These will require updating as the company grows and as legal requirements change.

Of course it's pointless if these policies aren't available to all staff on all projects, because it's important that employees understand and comply with them.

Operations manual

It's useful in bigger companies to have an operations manual which includes the company's policies and procedures. This manual can be used by all projects as a check-list to ensure that systems are applied uniformly and correctly. These manuals are particularly useful for new employees so that they can familiarise themselves with the company's method of doing business.

These manuals need to be updated regularly to ensure they remain current.

Standardised stationery

The company should have a standardised set of stationery and letterheads that are used on all projects. The company logo should be replicated from a master template with the correct size, proportions and colours. I've often seen companies use logos of varying styles creating confusion amongst clients and even looking unprofessional.

Where possible prepare standard templates for:

- daily diaries
- subcontract orders
- supplier orders
- invoices
- requests for information
- drawing registers
- drawing issue receipts
- day-works records
- site instructions or variation requests

Archiving documents

The following records should be kept and archived:

- personnel records, including:
    - records of pay and hours worked
    - disciplinary proceedings and warnings
    - employment contracts
    - medical reports
- safety records including:
    - accident reports and investigations
    - attendance records of inductions
    - the receipts of personal protective equipment
- tender submissions
- financial records including:
    - tax returns
    - payments made and their receipts
    - payments received and invoices
- plant and equipment service and repair records
- project records including:
    - correspondence to and from the client, the client's management team, subcontractors and suppliers
    - the final account and all measurements
    - variations
    - subcontractor documentation

- quality records
- drawings

Records must be stored where they won't be damaged by heat, rain, fire, insects or rodents. They should be easily retrievable and filed in date order.

The statutory period for which records must be retained is generally five years, but will vary between states and countries.

Self-perform or subcontract

Companies need to decide what work they will undertake themselves and what work will be given to subcontractors. This will dictate the number of employees and the skills that will be required.

The decision on whether to subcontract the work or not, will depend on the following:

- Can the contractor do the work more efficiently and cheaper than subcontractors?
- Are subcontractors available to carry out the work?
- Can the contractor recruit and retain the necessary skills for the work?
- Will the contractor have continuity of work for these new employees?
- Is the contractor able to purchase or hire the specialist equipment required for the work?
- If equipment is purchased can it be used on other projects?
- Some clients prefer using contractors who aren't dependent on subcontractors.
- Subcontractors may be unwilling to work in certain regions.
- Using subcontractors moves the risk away from the contractor but the contractor has less control over the work.

There are often specialist tasks that companies will subcontract because there will not be sufficient continuity of work for the company to be able to maintain the resources.

As the company grows and moves in different directions and into different fields they may choose to alter the model of operation and either subcontract more work or undertake more themselves.

Whether work is subcontracted or undertaken by the company may also depend on the individual project and its contract requirement. Sometimes, it may also be advantageous to use local contractors.

Reporting of problems & problem solving

In many companies there's a culture of hiding problems. This is often a result of managers blaming subordinates and implementing harsh measures against staff who may be responsible for the problem. It's important to foster a spirit of openness and ensure people understand it's not about blame, but rather about solving the problem, and being aware of a problem as soon as possible in order that mitigating measures can be implemented.

Of course where an individual has performed poorly, disregarded company policies and procedures or participated in fraudulent activities it will be necessary to institute the appropriate disciplinary procedures.

Regrettably many problems only come to light when it's too late to solve or mitigate them – possibly jeopardising the survival of the company.

When things go wrong

Unfortunately no matter how well projects are planned, or how well the company is run, problems will arise. It's essential that decisive action is taken to resolve the problem and prevent it from becoming more serious.

Contractors can be slow to solve a problem, fearful of incurring further costs. I'm not saying spend money at any cost, but often it's the only way to move forward. Normally the first thing contractors do when there's a financial problem on a project is cut costs, which inevitably means reducing personnel and equipment, in many cases making the problem worse.

When a problem occurs it's important to find the root causes of the problem. Often problems are only investigated superficially and a cause is uncovered which might not be the true reason, but only a symptom of the problem, or, it may be only one of the causes. For instance, if a project is losing money and investigation finds that the labour is unproductive, it usually doesn't help to simply reduce the number of workers, rather, it's important to uncover the reason for this poor productivity which could be due to a number of causes, including poor management.

Clients should be informed when there's a problem that affects them, and kept up to date with the steps being taken to resolve the situation. Clients appreciate the communication and some may even be able to help. Clients that find out about problems when it's too late can be quite unforgiving. Contractors shouldn't be too proud to ask for help from the client or even to approach other contractors, although this should be done only as a last resort.

Occasionally problems occur when key staff become ill, leave the company or have to deal with their own issues. Suitable replacements have to be found quickly without disrupting other projects or other departments. I've seen companies rob one project of personnel to solve another's problem, only to end up with two problem contracts. Of course it's sometimes possible to move people between projects and I've also seen individuals grow when faced with new challenges. The moral of the story is to ensure that careful consideration is made of the circumstances and people to ensure that appropriate measures are taken to solve the problem.

Ask for advice

There are too many components in the construction world for one person to know everything. If you are unsure of anything, ask for advice. Advice is often freely given, and you don't have to take heed of it if you don't like it.

Sometimes, there are people within the organisation who are more familiar with the subject matter than you. You may be surprised at the knowledge that some of your employees have, and anyway, they would probably appreciate being asked their opinion, even if the information isn't utilised.

Other sources of information could be sales representatives, subcontractors, suppliers, trade associations and even the client and their representatives.

Of course, always consider where the advice comes from, since some people may have alternative agendas, or might not be as knowledgeable as they purport to be.

Use experts

It's sometimes necessary to obtain expert advice for more complex problems. This may be as simple as obtaining accounting solutions, getting legal or contractual advice, or it may be for specialist solutions to complex construction problems. Usually the costs for these experts are far outweighed by the benefits of their help. For instance, although I have twenty years' experience with concrete, I have, on a number of occasions, paid for experts to advise me, and their advice has often saved me thousands of dollars.

Mistakes must become lessons

Over the years, I've seen errors repeated time and again, often by the same Project Manager, but certainly within the company on other projects. It's important to learn from mistakes so that they don't occur again, enabling the company to improve its performance. Just as important is to learn from successes and to replicate them.

At the end of every project the project team should analyse the successes and failures and see how things could be done better in the future. This exercise shouldn't be about pointing blame or looking for excuses. When the team has come up with the list of successes and failures, the reason for them and how things could be improved, it may be worth distributing the findings to other Project Managers, Head Office departments and managers. This may take the form of a simple memo or it could be a meeting where the issues can be discussed.

It's also useful to ask the client for feedback at the end of the project. This could take the form of a standard questionnaire. Obviously any feedback is useless unless it's actually analysed and the useful information is used.

Subcontractor and supplier performance data base

Good suppliers and subcontractors are often hard to find. Poor suppliers can delay projects by delivering material late or providing inferior materials. Substandard subcontractors delay projects, do work of poor quality, adversely affect the safety on a project and generally harm the contractor's reputation. It's therefore essential to look after good suppliers and subcontractors and ensure they are paid correctly and on time.

Contractors should keep a record and data base of all subcontractors and suppliers used. At the end of every project the Project Managers should rate them. Their performance should be assessed according to their:

- quality of work
- performance measured against the contract schedule
- safety

- responsiveness
- price
- likelihood of submitting claims

Further comments could include recommendations on:

- the type and size of project best suited to them
- the subcontractor's best team for future projects

These performance sheets must be collated into the data base which can be used by Estimators and Project Managers on future projects.

Be adaptive – adapt to changes in the environment

Construction is a changing environment and what's suitable today may not be suitable in the future. In fact what's suitable on one project might not be suitable on the next. Managers need to be able to adapt and change their approach between projects and with time. In fact, as the company grows so is it necessary for managers to change the way they do business.

Using technology

The implementation of new technology can often benefit the company, enabling it to be more efficient and productive. However:

- the technology must be appropriate to the company and their projects and care should be taken implementing systems and technology that hasn't specifically been developed or modified for the construction industry
- the technology must have suitable local back-up and support for both maintenance and training
- it should be simple to use
- personnel must be trained in the efficient use of the systems
- staff must be convinced of the suitability and safety of the systems
- it must be reliable
- the new technology must improve the company's operations, making it worthwhile to implement
- where necessary it must be able to integrate with existing systems and be compatible with the current popular software packages
- the technology must be adaptable and flexible
- the suppliers should be continuously evolving the system to keep up with technological improvements
- if necessary it should be able to be expanded
- the technology must be robust and able to work in remote and hostile environments

There are many different systems available and their prices vary enormously. Therefore, before deciding on a system it's important to adequately research the various options, decide what you require from the system, look at where the company will be in a few years' time (size, location and type of projects), and consider the pros and cons of each system and how they'll best suit the needs of the company in the future.

Implementing new systems

As the company grows it will be necessary to implement new systems. Many of the systems implemented by companies end in failure because:

- a system is selected which may be either unsuitable or too complicated and time consuming to operate
- managers haven't convinced staff of the reasons and advantages of the new system
- personnel haven't been trained to use it
- personnel are not given a chance to use the system immediately after the training meaning that they forget what they were shown
- personnel aren't encouraged and supported to use the new system which means that they fall back on the old system
- managers accept the continued use of the old system

Information Technology

Computers and systems must be adequately sized for the company. They must:

- not be slow
- have sufficient capacity to handle all of the company's tasks
- must allow for expansion
- be supported by local technicians
- support the software the company uses
- be compatible with the available technology

Most projects use computers extensively and it's important to ensure that staff have computers with sufficient capacity to meet the requirements of their duties. They may require specialised software, such as planning packages, which can be expensive, and should be managed to ensure that only products required for the particular project are loaded on their computers. Certain software packages can be helpful, and often their cost is far less than the benefits gained from using them.

It's important to regularly review software licenses. Often companies are paying for licenses which individuals no longer require. Many software suppliers also take a dim view when their products are used without the correct licensing and permissions. Often, as companies grow and more people use the software, the company forgets to buy additional licenses which could lead to them paying large penalties.

As the company grows it's beneficial to link all staff to a computer network, which may even be linked to the various projects enabling faster communication and the sharing of documents. These networks can be expensive to establish but there can be cost savings as well as improved efficiencies.

Computers, and the computer networks, must be protected from viral attacks, and data must be regularly backed up to a remote storage facility.

There should be policies and procedures in place governing the use of company computers which all staff must be aware of.

Unauthorised usage of computers:

- wastes time
- overloads and slows down the network
- leads to computer viruses
- may lead to inappropriate and sometimes even offensive use

Grand ideas

Sometimes managers and business owners have grand ideas about how to improve the company and increase profits. Sometimes these may be good but it's important to ask the following:

- Is it necessary?
- Will it make money?
- Does the company have suitable people to implement the idea?
- Will the results be worth the effort?
- Will it divert resources away from other areas in the company?
- Is this really about the good of the company or is it about benefitting someone's ego?
- Is the time right?
- Can it actually work or will it end up as a failure?
- Can we improve on what we are doing, with similar results but less effort, by implementing something else?

Of course if the answer to most of these questions justifies implementing the idea then it's important to explain to staff the reasons and steps required to implement the plan so that there's buy in.

Always look at new systems and organisational structures with a critical eye and ensure that the company has considered all the pros and the cons, taking into account all the costs, the impact of implementing the changes, and more importantly look at existing systems and structures and understand why they haven't worked or how they could be improved to maximise their benefits.

I've seen companies decide that a system, plan or division isn't working, in some cases even losing money. In an attempt to rectify matters they implement wholesale change, often implementing completely new systems and processes. In many cases due consideration hasn't been given as to why the system failed and how to rectify the problem. Instead, everything is changed, which often includes things which were good, or had the potential to be good.

I've also seen companies spend money on hiring consultants, having seminars, strategy sessions and trying to reinvent themselves with new logos, vision statements and catch phrases. Many of these have ended up as a waste of money as plans implemented were often superficial and didn't focus on the real problems of the company. In many cases staff wasn't committed to the changes, the managers eventually ran out of steam and enthusiasm, and all the work was for nothing.

I'm not saying companies shouldn't change. Of course they should. All companies can improve on the way they do business. The change must however be well considered, focussing on matters that will really make a difference. Often only small changes can be easily implemented with dramatic positive results.

Problems often occur because companies fail to do the basics right. Therefore, always make certain that the first principles are carried out properly, such as:

- projects are delivered on time, meeting the correct specifications and quality requirements and that they are done safely
- money due to the company is invoiced, and paid on time
- projects operate efficiently and avoid wastage of materials, people and equipment
- assets are looked after
- tenders are done correctly
- good clients are looked after

Change can be good, but it needs to be implemented for the right reasons and properly, and it's usually unnecessary to change something that isn't broken. However, most systems and businesses can be improved, which might just require minor changes and adjustments.

Don't get sentimental

Managers can't afford to be sentimental. This can be difficult, particularly for business owners that have built up their business from scratch. They're often attached to the first pieces of equipment that they bought, their first clients, the bank that first loaned them money or personnel that helped build the company. As the company grows and evolves it will be necessary to make some hard choices and move on, which may mean selling old equipment or letting go of people who may have become friends. Obviously with people and clients this should be done with care and compassion explaining to them why the changes are necessary and that it isn't personal but about growing the company and moving it forward for the benefit of other employees.

Case study:

One company had a Supervisor who had lost interest in work. He upset clients, subcontractors and the workers. On most days he did almost no work. Despite a number of us complaining to the owner the Supervisor remained employed. The owner firstly defended the Supervisor. When he couldn't defend him any longer he made excuses for not dismissing him, saying that he had to do it personally since he had known the Supervisor for so long. Yet, the owner was never available to talk to the Supervisor and when he was he forgot. This went on for months. The Supervisor earned a salary while the rest of the staff became exasperated. It also affected discipline and staff began to joke that you couldn't be dismissed from the company no matter what you did.

It would have been far better to have got rid of the Supervisor when it was apparent he was a problem. If the owner was so attached to him the money paid on the Supervisor's salary could have been better spent on giving him a big departure bonus, and everyone would have been far better off.

Plant and equipment

Companies often own assets such as earthmoving equipment, small power tools, vehicles, portable offices and computers. Many companies supply this equipment to projects at no charge. This can be a problem because:

- the true cost of constructing the project isn't reflected and the contractor has a false sense of what the work actually cost
- projects often don't use the equipment efficiently when they're not paying for it
- projects have no incentive to release the equipment to other sites
- Estimators tender for projects assuming that the item is available at no cost from within the company, then, when it isn't it has to be hired externally at a cost not budgeted for
- there's no way of working out the cost benefit of owning the equipment

Sometimes projects purchase the items which can also be problematic:

- items are purchased which are only useful for that particular project and cannot be used elsewhere in the company when the project is finished
- at the end of the project equipment is mislaid, sometimes ending up in someone's garage, on the rubbish tip or sold for less than it's worth
- equipment is bought which isn't compatible with other items owned by the company
- the item is used inefficiently and stands unused on the project for much of the time
- guarantees and warranties are lost
- the equipment may not be in the company colours with the correct logos
- the company ends up with a surplus of some items of equipment
- the equipment isn't maintained correctly

To ensure the efficient use and correct charge out for equipment I would recommend that equipment (other than small hand tools and portable electric tools) and buildings are bought by the company, then hired to the projects as required.

The hire rates should be calculated so that they:

- take into account the purchase cost spread over the life of the item
- allow for the estimated resale value of the item when it's disposed of
- allow for the cost of repairs and maintenance (excluding costs which are the responsibility of the project)
- are market related
- include for the insurance of the item
- allow for the finance costs

When purchasing equipment consider the following:

- will the internal hire cost be market related (we sometimes found for some items we couldn't compete with the price external hire companies charged so it was better to hire externally)
- are spare parts and servicing readily available
- the duration of the manufacture's guarantee and warranty periods and what they cover
- is there a guaranteed buy-back from the supplier

- is there a long term need for the item in the company
- are the items readily available from external hire companies
- is there an advantage to owning the item
- does the company have the resources and skills to maintain the item
- can the company afford the item
- could the money or finance be better utilised for other equipment
- is the item compatible with equipment the company already owns
- compare the total lifecycle costs of the various available makes and models (often cheaper models may be more expensive to maintain, use more fuel, be less efficient or require servicing more often)

Sometimes it's necessary to purchase equipment when:

- it's cheaper to own the item than to hire it externally
- it cannot be hired externally
- it will give the company a competitive advantage
- a useful item owned by the company becomes old and unreliable and needs to be replaced
- a project is in a remote region and it's better to have new equipment which usually requires less maintenance

There are other options to outright purchasing equipment such as leasing, or purchasing with a guaranteed buy-back. All options should be investigated and advice should be sought to see which is best suited to the company at that particular time taking into account any tax advantages and the financial capabilities of the company.

Reassess equipment

Companies that own their own equipment should assess it regularly. Old equipment not only portrays a poor image of the company but often breaks down, causing frustration on projects, disrupting the schedule and negatively impacting the productivity and becoming expensive to maintain.

New equipment:

- requires less maintenance
- is often under guarantee which covers some repair costs
- is usually more fuel efficient
- is more reliable and doesn't break down disrupting work
- is often more ergonomically comfortable for operators, reducing fatigue and improving productivity
- can be more efficient
- is often more powerful and can carry bigger loads
- is usually safer
- is often more environmentally friendly with less noise, exhaust discharge and oil leaks
- gives personnel a feeling of pride in their machines and the company
- is often more likely to be better looked after than older machines

In addition to assessing the age and reliability of the equipment look at its suitability. As the company grows the types of projects change which may result in some equipment becoming unsuitable or unnecessary and the items stand idle or are used for inappropriate tasks.

Although the costs of purchasing new equipment are large, the benefits often outweigh the costs. Alternatively, consider selling unsuitable equipment and externally hiring the correct equipment.

It can be beneficial to sell an item of equipment before it reaches the end of its useful life. As equipment gets older its maintenance costs increase and the price someone will pay for it decreases. There's usually an optimum time to dispose of equipment. One company I worked for kept accurate records of the purchase and maintenance costs for its equipment and consequently they were able to work out the best time to dispose of it. This varied between different types, makes and models of equipment. Obviously this is a guide and will vary, as sometimes even new items give problems from the beginning while others seem to work forever without problem. Some of this depends on the operator, the conditions the machine is working in, what it's used for and how well it's been maintained. It's therefore important to continually monitor breakdowns, repair costs and the hours worked of all major items of equipment so that appropriate timely decisions can be made to dispose of them.

Standardisation

As companies grow they will purchase equipment. Small companies with limited money and access to finance usually purchase the cheapest equipment, sometimes even second-hand equipment or end of range specials. Some owners put little thought into whether they need the item but buy it simply because it appears to be a bargain. This can result in the company owning a variety of equipment, including some brands that might now be unavailable, while others may no longer have spare parts readily available.

Different types of scaffolding and formwork equipment may be incompatible with each other. Often these differences are not always obvious (like the difference between imperial and metric systems) which may mean that the problem of incompatibility is only noticed after the equipment has been delivered to site which results in disruption and additional costs.

It's therefore important that early in the life of the company, systems and brands are chosen which are reputable, with parts readily available into the foreseeable future. This isn't to say that you won't find better brands later and switch to these.

We also found that certain suppliers were better at manufacturing certain types of equipment and often, say, bought excavators from one supplier and graders from another.

Reviewing the release of resources

Poor utilisation of resources can seriously affect the profitability of a project and the company. Not only is there the cost of wages, but the person or item could have been earning income elsewhere. It's vital to constantly review and update the resources that will be released from projects, and to allocate these to other projects.

To do this correctly projects must provide the accurate release dates as far in advance as possible, updating these requirements on a regular basis. Project Managers must be aware of where their released resources will be going, and the date they're required. Failure to release resources on the committed dates may negatively impact the next project.

Many Project Managers think only of the profitability of their own projects and not of the success of the business as a whole. Sometimes a Project Manager has to realise that releasing a resource earlier than they would have liked, or waiting a bit longer for equipment or an individual to become available, may impact their project, but it could benefit the company more when another project can better use the person or item of equipment. It may also mean that additional people don't have to be employed by the company, or new equipment bought or hired externally.

Management needs to ensure that the company's resources are used effectively and continuously so that their productivity is maximised.

Visiting projects

Managers should visit projects regularly because:

- project staff appreciate seeing senior management
- the client appreciates it when the contractor's senior management is interested in their project
- it provides an opportunity to see first-hand how the project is going
- they may see problems which others have missed
- they can pass on some of their knowledge and experience to project personnel

Visits shouldn't be rushed, allowing quality time to be spent with Project Managers. Often managers spend only a few hours on site, criticise the Project Manager, and leave. What has been achieved? Very little! The Project Manager hasn't been told how to correct the problems and no assistance has been suggested or implemented to help improve the situation. In fact, the Project Manager probably takes the criticism personally, becoming defensive instead of asking for solutions.

When planning the visit, managers should take cognisance of the Project Manager's commitments and plans for the day. I've seen senior managers visit on a day when the Project Manager isn't present (they're on leave, attending a course or visiting the client's office), they're tied up in meetings or busy completing month-end reports.

Obviously when Project Managers are absent for a period of time it's necessary for a manager to visit the project to ensure the works are proceeding without problems in the Project Manager's absence.

While I realise many senior managers have busy schedules they do have to consider why they're visiting the project and what they hope to achieve. In many instances it may be pointless visiting the project if the Project Manager is unavailable.

It's easy to visit a project, spend a few hours on the project and return home without achieving much. I found as a manager I needed to set goals and outcomes for my site visits to ensure I left with a better understanding of the project and that I made a meaningful contribution to the running of the project. I used to set a target to find a certain number of safety and quality issues as well as ways to improve the running of the project. Of course it shouldn't appear that you are only on a fault-finding mission.

Managers visiting the project should meet as many staff as possible. It's easy to not recognise staff or to forget their names. Sometimes there are new people you've never met. Consequently I always requested my Project Managers to introduce me to staff by name (even if I'd met them before) as we moved about the project. This way there was less chance of me embarrassing myself by not remembering a person's name or job title. I worked for one company where the CEO regularly called staff by the wrong name. This isn't really a poor reflection on the CEO as there were a few thousand staff, many of whom the CEO met only once or twice. However, damage is done when a manager walks past staff without greeting them or uses the incorrect name. A personal greeting from a senior manager, especially if there's an enquiry about a family member or a comment about previous successes, is sometimes as good for a person's morale as if they were given a pay increase. It's even better if the manager compliments them on a job well done.

Managers need to leave the project with a good understanding of the project's:

- problems
- progress measured against the schedule
- profitability
- safety
- quality
- resources
- outstanding payments
- unresolved claims and variations
- concerns which the client may have

Where necessary they should have provided advice on what corrective action needs to be implemented to support and assist the project.

Security

Theft can be a major cause of loss in a company. It can take the form of petty theft such as stationery, even larger items such as computers, and ranging to the theft of vehicles and equipment. Not only is there the physical cost to replace items but there's the additional costs of reporting and investigating the incidents, and more importantly, there's the disruption of day-to-day business caused by the missing materials or equipment.

It's also important to ensure that sensitive documentation is kept secured. This documentation would include:

- personnel records

- employees' rates of pay and bonuses
- records of disciplinary procedures
- personal evaluations and appraisals
- tender documentation
- project cost reports
- company financial records
- details of bank accounts and passwords

Sensitive documentation falling into the wrong hands could:

- jeopardise tenders
- damage client relations
- impact the outcome of negotiations over claims
- influence relationships with suppliers and subcontractors
- affect the company's ability to negotiate with clients, subcontractors and suppliers
- jeopardise industrial relations causing unhappiness and affecting morale and productivity
- result in poor press and bad publicity
- impact any legal proceedings and claims that the company may be involved with

Furthermore when documentation goes missing and there isn't a copy, time is wasted trying to recreate it.

To minimise the risk of theft ensure that:

- offices are locked after hours
- confidential documents are password protected on computer networks
- documentation is regularly backed up to a secure location
- sensitive documents are locked away when not in use
- only authorised personnel have access to confidential information
- personnel using this documentation are aware that it shouldn't be disseminated or left where unauthorised personnel may have access to it
- unauthorised people aren't permitted access to offices and work areas
- people working with confidential documentation may need to work in a secure office
- suitable security measures are implemented such as:
    - providing sufficient lighting
    - limiting entry and egress points at offices and project sites
    - fitting and arming alarm systems
    - fitting suitable locks, security doors and gates as required
    - installing fencing
    - hiring security guards as required
- ensuring vehicles are locked when not in use and the keys are stored in a secure place
- ensuring that personal computers aren't left in parked cars or open offices where they can be stolen

Many security measures cost money to implement, but their cost must be weighed against the risk and consequences of a theft occurring. However, many actions outlined above can readily be implemented at no cost and are simply a matter of people being aware of the likelihood and impact of a theft occurring, and ensuring that simple procedures are followed, such as locking office doors and windows before leaving.

Company store and yard

Many companies have a store, or even a yard. Some of these can be quite large. Unfortunately many of them aren't controlled properly.

Some important points regarding the store or yard are:

- they shouldn't become a dumping ground for the junk left over at the end of each project and only material that can be used on another project should be sent to the yard
- they should be kept neat and tidy and have proper stacking and storing areas
- they must be secure and controlled to prevent theft
- items that can be damaged by water or dust must be secured under cover
- the area must have adequate drainage
- they must be planned to allow access for deliveries and collections
- there should be separate, ventilated areas, with bunds, for hazardous and flammable liquids
- the correct permits must be in place
- sufficient firefighting and first-aid equipment must be available which should be inspected regularly
- they should be controlled and materials removed and received must be logged
- Project Managers must be aware of what's available in the yard

Manage risk

Contractors must manage risk. All projects have risks of varying degree so it's important to understand the type, level and amount of risk and ensure that:

- they are understood by everyone so that they can be managed and mitigating steps put in place
- if the risk eventuates the quantum isn't so enormous that it destroys the company
- there are suitable rewards for taking on the risks (add additional profit to high risk projects)
- where possible insurance is taken against the risk occurring (either externally or internally)
- if possible the risk is passed back to clients
- if it seems that the risk will eventuate, staff take immediate action to limit the impact to the project and company

Managing through boom periods

Sometimes contractors are lucky and there's so much construction work that they can take on as much work as they want at higher profits than normal. However, managing in a boom is often more challenging than operating in normal times, or even in a downturn.

Often there are so many opportunities that it's difficult to turn them away, resulting in the company taking on more and more work. Resources, both on the projects as well as in the office, become over extended. The company has to employ new people, often from a shrinking pool of available resources. At times the quality of people employed during these times is questionable. Management becomes stretched due to the additional projects, meaning that they have less time to spend with the new employees who don't understand the company's culture or the required quality and safety standards.

Furthermore, during a boom period profit margins on projects increase resulting in staff becoming complacent as it appears easy to make money. I've been involved with many projects where the tendered profit margin was 5% and we ended up making a profit of over 12%. On the other hand I've had projects with a tender mark-up of 20 or 25% which have seldom made the tendered profit, in many cases falling well short.

After undertaking projects in a boom it's difficult to get staff used to working in an environment of low profit margins where it's necessary to fight for every dollar to make any money. They've become used to making large profits even though they were wasteful on site.

When there's surplus of work available contractors also take on larger projects, some of which may be too big for the company, stretching its cash flow and resources, often resulting in the company going out of business or ruining its reputation. In fact, I know of many companies that have ruined a good reputation during times of abundant work.

Another problem is that suppliers and subcontractors also become busy, resulting in them raising their prices, which cuts into the contractor's profits. In addition, they have so much work it often results in poor service and longer lead times which again impacts on the profitability of the contractor.

Because there's so much work to be priced the tender departments become over extended resulting in them rushing tenders and making errors. Since the margin is higher the company becomes careless about analysing the risks. In fact, sometimes the company is so fixated about the seemingly wonderful profit that they will abandon all caution and good tender practices just to secure the project.

The important aspects of contracting in a boom period are not to become greedy, lazy or complacent, and be selective with what projects are tendered for. Try and assist your long term clients where possible as you'll need them when the market turns down, however, do take the opportunity to tender for larger projects, possibly in the next level above where you would normally tender, so that you can add these bigger projects into the company's portfolio.

It's important that during a boom period the company considers the downturn which will inevitably eventuate. This means ensuring that they build up sufficient capital reserves, don't take on more debt than they have to, or purchase equipment that'll be difficult to use during a downturn. When things are going well it's difficult to remember the bad times and imagine that they could return.

It's also wise to remember that additional people employed in the good times are going to need projects to keep them busy in the downturn, or they're going to have to be retrenched which could be a costly exercise.

Managing through a downturn

Starting a company during a downturn in the construction industry is often not as bad as it might sound. It gives the company a chance to grow slowly, developing robust systems that should stand it in good stead. The company usually also has the opportunity to find good skilled people from a large pool of available resources.

However, if the contractor has just been through a boom then a downturn can be stressful since management has to try and keep all their people, plant and equipment employed, or alternatively, they have to retrench employees and sell plant and equipment. It's important though, that the company doesn't take on projects that will lose money in an attempt to keep people employed.

A downturn should be used to trim dead wood in the company by retrenching those who are poor performers. However, where possible skilled people must be retained as they will be needed when the next boom comes. In fact, skilled people are even more important in a downturn because the company cannot afford errors on their projects and every dollar becomes critical.

Since suppliers and subcontractors will probably also be short of work it's imperative to negotiate better rates and payment deals. Of course they shouldn't be squeezed to such an extent that they're going to lose money on the deal and go out of business, or that they start taking short cuts or don't deliver the required service.

Everyone in the organisation needs to understand how important it is to find work. Project Managers should try and eke out as much extra work as possible on their existing projects, no matter how small. Not only will this help to keep resources employed on the project longer, but, often these projects have been tendered with a higher margin than current tenders so it could be more profitable to do bits and pieces on the existing projects. Where possible, encourage and help the client by highlighting opportunities for additional work.

A major problem is when the downturn is felt across the whole economy and it results in clients experiencing financial difficulties and being unable to pay the contractor for completed work. This impacts the contractor's cash flow and can easily send them out of business.

It's easy to panic in a downturn, making rash business decisions or submitting risky tenders, which often only makes the company's position worse. In all of this, communicate with staff since many of them will start to feel insecure, resulting in them becoming demotivated and even leaving the company. Unfortunately, it's usually the better employees that leave first.

If the company is in trouble it's often good practice to be honest with clients and ask them to assist by paying a few days earlier or even awarding the company some additional work. It's not in the clients' interests to lose a good contractor, and especially one that's in the middle of one of their contracts. Of course staff shouldn't know if the company is experiencing financial problems as this could result in employees leaving in panic.

Corporate social investment

Most companies will be regularly approached by charities, local organisations, schools, and so on for donations. Nearly all companies make some form of contribution which may just be a few dollars, several thousand dollars, or in some cases isn't a cash donation but is a contribution of time and machines. Most companies don't track these donations and would be quite surprised at the total amount, especially when the contributions are added in from the different projects.

In general I would recommend that all the donations and contributions of money, time and equipment are recorded since the value of the contributions, particularly those to local communities, can be a useful marketing tool. As discussed in Chapter 2, if possible try and rather contribute a few larger amounts than many smaller amounts, and if possible contribute towards particular items, like bursaries, new computers or new classrooms, rather than just a general cash donation to the organisation. The handing over of these donations or awarding of the bursaries can be used as an opportunity to advertise the company by inviting the press and clients. Obviously the contribution must be meaningful and worthwhile, otherwise it can seem ridiculous.

While a company is successful they should be contributing to good causes, particularly if the success has been due to working in a particular area or community. However, if the company is going through a poor financial period it may be necessary to cut back on donations to ensure the company is able to survive through the difficult times.

Summary

- To successfully manage a construction company managers require many attributes and knowledge of the local laws and legislation. They should also:
    - have an open door policy
    - avoid implementing systems that waste time
    - limit the number and length of meetings and ensure that only those relevant to the meeting attend
    - ensure that important tasks are done on time
    - learn to say no
    - be able to communicate, persuade and negotiate
    - stand up for their team when they're right
    - make timely decisions taking into account the relevant available information
- Companies must have appropriate structures and reporting lines.
- Unnecessary company overheads must be avoided and Head Office running costs must be limited.
- Appropriate departments may be set up to control some of the company's functions.
- As companies grow it may be necessary to create different divisions responsible for different types of work or different areas, however, these divisions shouldn't operate completely independently of each other and

must be willing to cooperate and share resources.

- Managers must take an active role in promoting and monitoring safety.
- Tender systems should be implemented which simplify the tender process, make it more reliable, providing useful information which can be included in the submission and complement the running of the project.
- Insurances must be reviewed regularly.
- The company must have valid licenses, permits and registrations to operate in each area and field of operation.
- Guarantees must be retrieved from clients as soon as practical.
- To enable the company to operate efficiently:
    - suitable document control systems must be implemented
    - policies and procedures must be in place and be available and understood by staff
    - an operations manual should be maintained
    - standardised stationery should be prepared and available
    - documents must be correctly archived in a safe location
- The company must decide what work they will do in-house or self-perform and what they will subcontract.
- It's important that problems are reported, mitigated and solved as soon as possible.
- When unsure of something managers should seek advice and even engage experts to provide the solution.
- Mistakes must become lessons.
- It's useful to maintain a data base of subcontractors and suppliers which records their performance.
- Managers need to adapt to changes in the environment.
- Technology can be useful but new technology must be implemented properly and needs to be suitable for the needs of the company.
- New systems are sometimes necessary but care needs to be taken to ensure that they are appropriate and are implemented correctly.
- Sometimes change is required, but managers must resist implementing grand ideas which are unnecessary, inappropriate and costly. Minor adjustments focussing on the basics could be far more effective, incur fewer costs and result in fewer disruptions.
- Managers shouldn't be sentimental and must make decisions based on sound business principles.
- Plant and equipment owned by the company should be:
    - managed properly so that it's productive
    - hired to the projects
    - assessed regularly to ensure that it's appropriate for the work being undertaken by the company
    - replaced when necessary
    - standardised where possible
- Resources must be regularly reviewed to ensure that they are productive, are not idle and are released timeously for other projects.
- Visiting projects is an important aspect of a manager's duties and these

visits must be coordinated so that sufficient time is spent with Project Managers enabling value to be added to the project and so that managers leave with a full understanding of the problems, risks and profitability of the project.

- It's vital that steps are implemented to ensure the security of property and information.
- Company stores and yards must be properly managed and maintained.
- Risks must be managed and where possible avoided, or measures taken to mitigate them.
- Appropriate steps must be taken to successfully manage the company both in a boom period and during a downturn.
- Corporate social investment is important, but it must be appropriate, relevant to the company's operations and tracked so that the total value is known.

Chapter 12 – Growing the Company

Many contractors start off successfully. They develop a good reputation and are profitable. The company grows. Then the cracks appear. Quality deteriorates. Projects are no longer profitable. The company starts losing money.

Why does this happen? Well there are a number of reasons.

- Normally a company starts off small and the owner is hands-on, full-time supervising projects. When the company grows the owner can no longer be hands-on and passes control to managers and staff who aren't necessarily as focused on ensuring that the projects are successful, and don't take the same pride in looking after clients and achieving quality work as the owner did.
- Sometimes the new managers are inexperienced. Small businesses often employ mediocre staff because they're unable to pay large salaries or attract good people.
- Often owners don't delegate work correctly. They try and remain in control, but have too much other work and are unable to attend to everything, leaving projects waiting for materials, resources, decisions and guidance.
- Clients become frustrated when the company isn't responsive because the owner is busy attending to other business.
- In addition, the company undertakes more and bigger projects, resulting in them having insufficient cash flow. Lack of cash disrupts the procurement of materials and payment of subcontractors, causing delays and, in the worst case, the company goes insolvent.
- Sometimes the company experiences a rapid expansion in a boom period and employs more staff, moves into bigger offices and purchases equipment. Unfortunately these boom periods are short lived and end just as suddenly as they started. The contractor is left with excess staff and equipment in offices that are too large with a high rental.

However, it's important that businesses grow because it enables staff to develop and creates opportunities for advancement. It's also the only way that business owners can move out of managing projects to managing a company. However the company must grow in a sustainable manner and growth needs to be managed without harming the company's reputation.

Controlled growth

It's imperative that the company grows in a controlled manner. The owner must gradually hand over control of various areas of the business to trusted, experienced individuals. The company must progressively increase the number and size of the projects they undertake, ensuring that they always have sufficient cash to finance their operations (including maintaining a buffer for when the unexpected happens) and sufficient resources to carry out the work.

Is there a long term future?

Many contractors launch themselves into new ventures in different fields and in new areas, often purchasing equipment and employing personnel without considering the long term future and viability of the venture.

Case study:

One company decided to start a core drilling and concrete cutting division, so they purchased a vehicle with specialised equipment. However, they hadn't researched the viability of the idea, nor prepared a financial feasibility study.

There was insufficient work in the region to keep a team fully occupied, and certainly not enough work to justify the capital expenditure on the equipment. Furthermore, no one was allocated the task of looking after the new business and marketing it. Consequently the vehicle and equipment stood idle and slowly items were removed to use elsewhere. Eventually it was no longer a useful or productive unit.

Many projects and business ventures seem to be a good idea at first glance, yet it's important to first perform a feasibility study to check their viability and potential profitability. It's equally important to check the long term sustainability of the venture since many products or services can be in high demand, but their demand is short lived.

Not haphazard

Many companies grow in a haphazard fashion without planning for the future needs of the organisation. People are employed to fill vacancies with little thought as to whether the position will be required in the future, or whether the candidate will be suitable for that position in a bigger company.

Departments are formed and grown haphazardly which results in them having to be restructured, or even done away with, as the company grows. Reporting structures or managers may have to be changed to suit the larger company.

I'm not saying that the company must now set up all the departments and structures that'll be required in five years' time, nonetheless they should at least have an idea of what the structure will look like in a few years. The plan will evolve and change with time, but the important part is that the current structures should also be able to evolve and change.

Some companies also target a varied type of work, employing skills and purchasing equipment for one project which maybe can't be used on the next, with little idea of the direction and type of work they want to target in the future.

Not growth at any cost

Some business owners and managers are driven to grow the business no matter what the conditions are. Large companies have shareholders who expect the company revenue to increase every year, particularly after the company has experienced a few years of stellar growth. Unfortunately construction has periods of growth and times when work is in short supply. It's inadvisable to try and expand the company in times when work is limited.

Of course some managers are incentivised by pay increases and big bonuses to drive the growth. As a result they often do this at any cost, with little thought for the future of the company, purely focussed on the short term results to satisfy their personal bank balances and egos.

Systems

It's difficult for a company to grow if it doesn't have appropriate systems in place. These systems include financial (to handle debtors and creditors), cost control, payroll and tendering. They need to be simple to operate but allow for future expansion.

Manual systems can be useful for small businesses but rapidly become inadequate when the business grows.

I often see large companies running with systems which are more suited to small businesses. These systems cannot handle the volume of information, resulting in disruptions, causing the company to eventually migrate to another system. This causes further inconvenience since personnel have to be trained to operate the new system and existing data has to be transferred (which takes time and may result in data being lost).

People

Growth can't be achieved without the right people. Unfortunately many smaller companies employ people used to working in a small company, who don't have experience and knowledge of how large companies operate. This could be because when the business started the owner employed friends, acquaintances or relatives. Also there are usually limitations on salaries a small company can afford to pay, and in any event, people who've previously worked for a large contractor aren't easily attracted to a small company.

Eventually companies can outgrow the people originally employed. Don't get me wrong, many people do grow and develop as the company grows, but unfortunately, some don't and ultimately become a burden. It eventually becomes necessary to ask them to leave or move to a position better suited to them. However, transferring a long term employee to another position, or bringing in someone from outside or from under them to become their manager, can result in an unhappy and unproductive employee so it needs to be done with care.

Cash flow and guarantees

Cash flow is a major stumbling block to the growth of a company. A company can be generating good profits but if this profit isn't turned into cash in the bank they won't be able to finance the growth of the company.

Bigger and more projects strain cash flow further since more money is tied up in retention and more is required to pay wages and suppliers.

Compounding the cash flow problems is that obtaining additional surety facilities for extra projects usually requires the contractor to deposit additional cash with the institution as a guarantee for them issuing the sureties and bonds.

Supporting departments

As the company grows it becomes necessary to employ more Head Office personnel to support the projects, often resulting in the creation of new departments. These may include; financial, tendering, quality, safety, planning, legal and contractual.

Of course these aren't all required straight away and will develop in stages as the company grows, but they add to the company's overhead costs.

Registrations for quality, environmental & safety international standards

As a company grows it will need to obtain the appropriate registration and certification that confirms that they comply with quality, safety and environmental standards. Certification is often a prerequisite to be able to tender for many projects. Even if it's not, it can form an important part in securing the award of a project.

The various certifications require the contractor to have systems in place to ensure that the company's operations are compliant with specific recognised standards. In addition compliance needs to be monitored since audits are performed regularly by outside agents to ensure standards are maintained. Setting up the systems and ensuring continued compliance costs time and effort. Much of the work can be done by external agencies or contractors, but ultimately those working within the company will need to operate the systems.

More projects or bigger projects

A company can grow by undertaking more projects, or larger ones. As discussed in Chapter 1 it's often simpler to have a few larger projects than several smaller ones. However, clients won't award a large contract to a contractor they consider doesn't have the experience or means to tackle it. Large projects also come with their own sets of risks as discussed earlier and contractors need to ensure they have the resources and financial capacity to undertake them.

Few contractors will be able to grow without undertaking larger projects, so the optimum is to grow by taking on more projects and a few larger ones as well.

Where to next – location, field, client.

Often growth potential is limited in the area, or in the field in which the company normally operates. To expand, the company may have to consider working in other regions, possibly forming relations with new clients and even developing new skills to work in other trades or fields.

The important part is not to let this new business distract the company from doing what they are doing well. Often I see new endeavours take an extraordinary amount of management time and company resources, resulting in the company neglecting their existing projects and clients, culminating in additional costs and even lost business.

Having said this though, it's important that a new area of business isn't tackled in a half-hearted manner causing failure. It's a fine balance to continue to manage the existing business while successfully expanding into new markets and regions.

It's usually best to stick to a field related to the one that the contractor is familiar with, which requires people with similar abilities and uses related equipment. I've seen many contractors branch out into completely different fields, with no similarities to the one they're currently working in, often ending in expensive failure because the company doesn't have the necessary expertise or experience. As the saying goes, 'stick to your knitting'. But, I would add, do try knitting different items.

Expanding to other states or countries

Many companies expand by working on projects in other states or countries. Several of these ventures are very successful, but, for every success story there's a horror story of companies losing millions or even going out of business.

Before operating in another country consider:

- Do you understand the legislation and rules in the country?
- What permits and registrations are required?
- Will the legal system protect the company if things go wrong?
- Will the company be paid?
- Do you have personnel willing to work in these countries?
- What are the logistics to transport goods and people to the project site?
- Are there diseases, particularly in tropical areas (like malaria), which could pose a risk to employees?
- What are the taxes and how are they implemented:
  - GST or VAT
  - import duties for materials and equipment
  - what is the tax rate on companies
  - is there a tax treaty between the countries so that tax isn't paid both in the country where the project is located and the company's home country
- Will you be able to get your equipment out of the country when the project is completed?
- Are there security risks?
- Have you considered the climate? For instance, tropical regions can be very wet at certain times of the year causing disruption and delays.
- Is the country politically stable?
- Are materials and equipment available and what are their costs?
- Is there a risk of major currency fluctuations?
- Will the company be able to take their money out of the country?
- Are the required skills available in the country?
- Will there be suitable medical treatment available to treat personnel who are injured or become ill on the project, or should emergency evacuation procedures and insurance be put in place?

- What are the costs and requirements for taking staff and workers into the country?
- Is the country generally free from corruption (working in countries where bribery is rife not only makes it expensive to do business but it's also dangerous, so they should be avoided whenever possible)?

Many of the above will also have to be considered when working in a different region within the same country. If the costs and risks have not been adequately considered during tender stage the project can turn into a disaster, ending up costing the company money.

Purchasing another company

One method of growing is to purchase another company. This, though, usually requires substantial funds and is fraught with risks. I worked for a company that purchased a number of companies over a period of time. Some of these purchases were very successful and contributed to the growth of the company. Unfortunately about half of the purchases were unsuccessful with a few turning out to be more costly than expected.

Many companies purchase a poorly performing company with the intention that they will be able to turn the company around. Regrettably this isn't always as easy as it seems. The new company often demands lots of management time which is diverted away from doing what they were successfully doing – which was running a good company. I've often seen the diversion of management adversely affecting the performance and reputation of the original company.

It's important to carefully analyse the reasons for purchasing the company. Only purchase a company if it's going to offer new opportunities. These may be because the new company:

- Is working in a different region and it will help your company expand into this region. Nevertheless, it's important to first:
  - analyse the risks and opportunities of the region
  - understand whether the new company is operating successfully in the region
  - check that they are making money
  - understand the reasons for their success
  - ensure that there will be further work in the region
  - understand whether the company will provide the appropriate spring-board to expand into the region
- Has established relationships with different clients to the ones you normally work with. However, you must first understand:
  - the reasons for these relationships since they are often built around personal relationships which can be worthless should individuals leave the client's or contractor's employ
  - whether the clients will continue to provide work opportunities
- Has expertise in a different field. Nevertheless, first check the following:
  - is this expertise because of particular individuals employed by the company, and will they remain with the company after the take-over

  - could these individuals, or others with similar expertise, be engaged by your company at a fraction of the cost of purchasing the whole company just to get their knowledge
  - is the whole company expert in this field, or are you purchasing other sections and people which are less useful and may even have to be terminated later at additional costs
  - is the company really an expert and is their reputation with clients as good as you perceive it to be
  - will having expertise in this field enhance the company's capabilities and profits
- Has a large pool of personnel. But first ascertain:
  - whether the personnel will remain with the company (I've generally found up to half the employees resign and leave the company within a couple of years of the buy-out – unfortunately it's often the better ones that leave first)
  - if it would have been cheaper to attract the skilled people from elsewhere, rather than purchasing the company to get the people
  - how many of the people aren't good and are dead-wood who will need to be made redundant
  - the overlap of resources between the companies since it is invariably necessary to terminate staff when there is already a person fulfilling that function (this termination process is expensive and causes fear and resentment amongst the new company's employees – leading to resignations and poor performance)
- Has lots of equipment. However, it's necessary to check:
  - if the equipment can be utilised (remember there's no choice and you're paying for all of the equipment whether you can use it or not)
  - are the items compatible with the equipment already owned
  - if the equipment is of a similar model and make as equipment already owned, since having assorted models and makes means additional spares have to be kept and maintenance is more difficult
  - the condition of the machines and have they been regularly maintained with service histories and records
  - are the warranties in place and still valid (because some may have been invalidated if the equipment hasn't been serviced correctly or the original manufacturer's parts haven't been used)
  - the condition of wearing parts and tyres as there could be a large cost to replace them
  - are there facilities and people in place that will continue to maintain the equipment
  - are spare parts readily available for all the items
  - who owns the equipment

- Alternatively by purchasing the company you may be removing a competitor. But is this:
    - going to be a benefit worth the cost
    - not going to result in another company taking their place

It's imperative that a proper due diligence is carried out on the new company. This should include reviewing:

- many of the items above
- the financial statements and verifying their accuracy
- all projects and understanding their safety, quality, outstanding payments, what costs and revenues have been declared and whether these are correct, potential problems, progress measured against the project schedule, forecast cost to completion, resources, suppliers and subcontractors, the status of claims and variations, if the client is happy (review project progress meetings) and the quality of staff
- outstanding debtors and creditors
- business systems
- employment contracts
- accident and insurance claims
- unresolved problems on past projects
- tenders which have been submitted and may be awarded
- operating procedures
- their tax affairs
- outstanding guarantees and warranties
- labour relations and agreements
- the numbers of staff, their skills and years of tenure
- lease and rental agreements
- money owing on equipment and other assets

Buying another company and integrating it with your company can take a number of routes which will depend on the region in which the company operates and its size and field of expertise. The company can:

- be left to operate as a separate division within the company
- be left to keep its name and continue to operate as a separate company
- operate as either of the two forms above but with key people from your company inserted into the new company's management structures
- be fully integrated and merged with an existing division, or the company as a whole, losing its identity completely
- part of the company may be integrated and part may operate separately, or parts may even be broken out and sold

When purchasing a new company it's important to have a plan in place to ensure that the integration happens speedily and efficiently. There will always be lots of uncertainty amongst the new company's staff as well as their clients and suppliers. This uncertainty may result in clients taking their business elsewhere, suppliers cutting off deliveries and staff resigning or operating at reduced productivity.

As part of the merge of the companies the following should happen:

- Before buying the new company have a clear vision of how the merge will take place and ensure that the senior management of the other company understands this vision.
- Once the purchase has happened, communicate the vision, processes and advantages of the purchase to staff of both companies. Senior management from both companies may have to talk to all staff (it should be noted that this is usually more of an overview rather than a detailed step-by-step view).
- Meet with key clients to inform them that it should be business as usual but outline the advantages to them of doing business with the bigger company.
- Meet with key suppliers, or send them a letter, to reassure them that accounts will be paid. If necessary provide them with new details of where their accounts should be addressed as well as highlighting any different processes.
- Ensure key staff have bought into the idea and are committed to make it work so that they communicate and convince other staff of the advantages and benefits.
- Take steps to protect and secure equipment because often staff-members take the opportunity to remove company property in the confusion created by the merger.

Cost of growth

Growth can be costly. As the company takes on more and bigger projects so to must the administrative and support systems and staff grow. The increase in these systems results in the company's overheads increasing. Unfortunately the growth in administration and support is often not directly proportional to the increase in turnover. For instance, the company may employ one person in the payroll department but when the number of employees becomes too much for one person to handle a second assistant must be employed. The turnover of the company hasn't doubled, yet the salaries in the payroll department have. Eventually the company may even have to move into bigger offices which results in more costs. However there will also be times when the growth in turnover doesn't result in a growth in overheads since better efficiencies are achieved with the same resources.

Sometimes growth costs money because:

- new equipment must be purchased
- new regional offices must be set up and staff employed for these areas
- there are often unforseen learning costs of operating in new areas or in a different field
- new people are employed with specialist knowledge and they may not earn revenue for the company for several months until they secure a project

- additional funds are required as security for more and bigger sureties and bonds

Contractors must consider all the additional costs, carefully weighing up whether the growth is sustainable and that the costs can be recouped and rewarded several times over.

Summary

Many successful contractors fail when they grow.

It's important that growth is:

- done in a controlled manner with owners gradually handing various areas of the business to trusted and knowledgeable individuals
- pursued in areas where there's a future for further work
- not haphazard and that the company has a vision for the future
- not pursued at any cost

To grow the company:

- requires suitable people
- there must be appropriate systems in place to support the growth
- there must be sufficient funds to finance the growth
- requires support departments and Head Office staff
- often requires the company register with and comply with recognised standardisation bodies

Growth can be achieved by:

- doing more or larger projects
- expanding to new locations
- developing new clients
- branching into new fields
- going into other countries and regions
- purchasing another construction company

Growing a company can be costly so it must be well considered. It's however essential for most companies to grow.

Chapter 13 – Reputation

Contractors are only as good as their last mistake. Unfortunately most people remember the faults and very few seem to remember the good work and successful projects. When asked which restaurants gave you good service or scrumptious food you'll probably have to think, but if you are asked which restaurant gave you bad service or poor food, you'll probably recall the name instantaneously, even remembering the date and who you were with at the time. Construction is much the same and clients will remember errors long after they've been remedied. They can be unforgiving. Even if your price is the cheapest but you have a poor reputation (which may even be undeserved) you will not be awarded the project.

One mistake can undo thousands of hours of hard work. That error may be caused by you or one of your employees. Even incidents that you consider trivial can be problematic for the client (sometimes it shouldn't be a problem but clients view things differently and have different priorities and agendas). Incidents that can affect a company's reputation include poor quality, ill-disciplined staff, aggressive behaviour when settling claims and variations, safety incidents, environmental accidents and not delivering a project on time.

Furthermore, clients talk to each other and word quickly spreads that a particular contractor is problematic.

It's important that all personnel understand the importance of delivering a quality project on time without safety, environmental or disciplinary breaches. Reputation is a team effort and everyone needs to understand how important it is for them to portray the company in a good light, maintaining the company's fine reputation in everything they do.

Proactive

Clients like to work with contractors who are proactive and anticipate potential problems so that they can be solved before they occur. It's important to work with the client as a team, communicating with them regularly, so that they are made aware of possible complications and variations as soon as they become apparent.

Assist the client with solutions to their problems. Remember that you often have more experience than most clients. In offering solutions, though, be cautious that you're not taking on the responsibility for the solution. It's important that the client's designers verify the proposed solution is suitable.

Continually advise the client of increases in costs, particularly if the client's budget could be exceeded. This will enable the client to either arrange the additional funding or cut costs elsewhere.

Responsive

Contractors need to be responsive to clients' demands. I don't mean do work without charge, but rather try and accommodate the client's changes and additions where possible. The nature of most projects is that the client will change the schedule, their milestones, as well altering structures, buildings and finishes. These changes can often be frustrating and take up the contractor's management time. It's frequently easier to say no to the client since the additional variations may barely cover the cost of doing the work. But, by always saying no the contractor will quickly get a reputation of being uncooperative.

Being responsive also means returning the client's phone messages, making yourself available for meetings, responding to queries promptly and submitting prices and revised schedules on time.

In addition, when there's a problem, either during construction or after the project is completed, it's essential that the contractor responds and rectifies the problem as soon as possible if it has been caused by them.

Being responsive however doesn't mean that you must accept being bullied by a client who doesn't respect the contractor's rights.

Fair-minded

Clients must perceive the contractor as being fair. This doesn't mean that the contractor shouldn't charge for additional work or variations, but rather that the contractor shouldn't take undue advantage of the client and charge excessively for these additions, or charge for work they haven't undertaken.

All of this is about perception, so it's important that the contractor highlights to the client items which they have carried out for a reduced rate or no charge, and to explain why the rate for some work seems to be excessively high.

When something has been taken out of the scope of the works the contractor should pass on the saving to the client.

Case study:

Many years ago a client awarded us a project worth ten million dollars. We had been on site for a few weeks busy with the excavations and site establishment when the client cancelled the contract because they had changed their plans. We filled in the excavations and demobilised from site, charging the client only for the work we had completed.

In terms of the contract we didn't have rights to claim for anything more than this, however the client believed that we could have charged them for the loss of profit on the cancelled work and they were relieved we had been so reasonable with our claim. Because of this they felt that they had some obligation to repay us, so over the next few years they awarded us several large projects which in aggregate were nearly three times bigger than the cancelled project.

How much profit is too much?

This sounds a stupid question and I'm sure I'll be ridiculed. Surely there can't be such a thing as too much profit? I've been fortunate to have completed a few projects where we've made substantial profits. However, the amount of profit does depend on the circumstances and the clients. It's important, though, that the client believes that they are getting a fair deal. If they don't they will simply take their business elsewhere.

To a lesser extent the amount of profit also depends on how much the client is prepared to pay, which may indicate I'm advocating richer clients should be charged more than other clients, which I'm not. It's just simply that when projects become too expensive they are no longer viable and the client won't proceed with them. Furthermore, a client who runs over budget on one project is often reluctant to start the next project.

Working with clients can sometimes be likened to having a cow. You can keep milking it every day, making some money from it for many years, or you can slaughter it and make lots of money once.

Some contractors get a reputation for overcharging clients, or being too expensive and clients simply stop using them.

Safety

It's essential that construction companies take safety seriously. I've known contractors being barred from working for certain clients because of their poor safety record.

Some clients pay their managers bonuses which depend on their safety record. Therefore these managers won't tolerate unacceptable safety practices from their contractors because it will adversely affect their bonus.

In addition, accidents create work. Incidents have to be investigated, reports written and explanations provided to management. Nobody likes additional work — especially the client's team! Serious accidents may even result in people losing their jobs. Consequently most clients will go out of their way to ensure that the contractor they pick has a good safety record.

Also, don't underestimate the poor publicity a company gets when there's a serious accident on one of their projects.

Quality

As mentioned quality workmanship and materials are essential, not only to prevent the company from incurring additional costs, but also so that the company has a reputation for delivering quality projects.

Most clients are prepared to pay a premium to obtain a quality product and efficient service since this:

- often reduces their supervision costs during the construction process because they know they can rely on the contractor to produce the required standards with minimal supervision
- causes fewer problems with their follow-on trades because items are more likely to fit, reducing the chance of delays or additional costs to the project
- usually means the product will last longer and require less maintenance
- looks good and adds value to many facilities (particularly houses and apartments which they are going to sell)

Managers should be proud of the quality of work that the company produces and encourage all employees to take pride in their work.

Professionalism

Always be professional and courteous in your dealings with the client. It's easy to say things in the heat of the moment which could upset clients. Some clients can be unforgiving and bear a grudge for years, jeopardising the chances of winning work from them.

Yes, I've personally had heated arguments with clients over issues where we've disagreed, but I've never held a grudge or said anything personal to them. In fact, some of them still delight in reminding me of some of our arguments.

Professionalism also extends to how you dress, that you're polite and respectful, appear organised, and arrive at meetings on time and prepared. It includes the way you are seen to treat your employees, suppliers, and subcontractors, and generally carry out business.

Well organised contractors who know what they are doing, have the right tools, people and equipment, and are properly organised on site with appropriate neat offices and facilities, all give the aura of professionalism.

Clients want to feel they are dealing with professionals.

Honesty and integrity

The company and its personnel must be seen to be honest. This is more than just not over-charging the client. It's about conducting yourself and the business beyond reproach.

Employees shouldn't be stealing from the client, fellow workers, subcontractors or neighbours. The contractor shouldn't be paying bribes and should be seen to treat their suppliers and subcontractors fairly. Not only don't clients want any added trouble, the disruption of police investigations or the ire of neighbours, but they will certainly wonder if the contractor condones this type of behaviour, what other illegal behaviour is occurring and whether they are being swindled by the contractor.

Be seen, be involved

The contractor's managers need to be seen by the client to be involved in projects.

When I've been in senior management I've always tried to visit my projects regularly, where possible taking time to attend the project progress meetings. These provided an opportunity to meet the client's team and to understand concerns they may have had. It was also a chance to afford support for my Project Managers.

I've often been complimented by clients for this visibility and was frequently told that they never saw some of our competitor's management.

Of course, when there are serious problems on the project it's even more important to visit the project regularly. It not only provides support for your team but it does seem to calm the clients, giving them assurance that their problem is being taken seriously by the contractor. In fact visiting the project often pre-empted irate telephone calls and letters. Clients are usually far more civil when you're standing in front of them discussing a problem.

Meeting the client's expectations

It's important that the contractor is able to meet the client's expectations. This means that they need to understand what these are, and furthermore ensure that they are reasonable and that the contractor can deliver on them. However, it may be necessary to discuss these expectations with the client, ensuring you both have the same understanding of what the finished product will look like and what it will cost.

Sometimes contractors have to explain why the items and finishes the client has chosen won't provide the final product they're expecting. Failure to manage their expectations, or to deliver on them, will certainly end with an unhappy client which could adversely affect the contractor's reputation and even lead to a dispute.

Delivering projects on time

Clients will often pay a premium to employ contractors they know will deliver a project on time. For many clients time is money. The sooner a project is completed and they can start operating the facility the sooner they can earn money, or the sooner they can move into a new house or office so they can stop renting their current property.

It's important to work with clients and develop a schedule which is achievable, but at the same time one which meets their needs and requirements.

Failure to meet the schedule is not only costly but damages the contractor's reputation. Large public projects which aren't delivered on time will even attract bad publicity for the contractor in the news media.

Do not over-promise and under-deliver

I've had managers and Project Managers who continually over-promised and have committed to dates and requirements which were impossible to meet. When these dates weren't met, or requirements weren't delivered, the client became unhappy, disillusioned with the contractor and eventually didn't believe anything the Project Manager said.

It's usually far better to under-promise and over-deliver. The client is then pleasantly surprised when the task is delivered ahead of time. Of course the client's not going to accept dates which are clearly extended further than they should be.

Wherever possible contractors must ensure that they achieve their commitments, providing the client with confidence in their ability and enhancing the company's reputation for being reliable.

Mistakes happen

Unfortunately mistakes happen on projects. As much as you've planned and staffed a project properly, and done everything correctly, problems will and do occur. When a problem arises the contractor is often judged on how the problem is resolved, rather than there being a problem in the first place.

Case study:

When we were renovating our house the builder installed the vanity cupboards in the bathroom on a Friday morning. We weren't happy with them and immediately emailed our concerns to the Project Manager. He responded saying he would be on site on Monday morning with the Supervisor. I expected there would be an argument and they would try and convince me the cupboards were acceptable. However, when they arrived they listened to why I felt the unit wasn't correct and agreed with me. On Friday the replacement cupboards arrived.

Whenever I recommend this builder to others I use this story – although the builder made a mistake, they responded quickly, accepted it was their fault and immediately rectified the problem, giving us a finished product we were happy with.

Of course, not all problems are necessarily due to the contractor. Sometimes, there are defects with the client's design or the materials they've specified, or at times, the client has unreasonable expectations or has misinterpreted a drawing or specification.

Therefore it's always essential to assess the cause of the fault. If it's due to the client it's important to explain to them what caused the problem, why it's not the contractor's fault and to suggest ways to rectify it. On occasion the contractor may need to engage expert opinion to support their argument. In all of this remember that time is usually of the essence, so the investigation and repairs need to be completed as soon as possible.

Commitment from employees

All employees are an advert for the company. I've had clients unhappy with the company and the type of people we employed because our workers were overheard making inappropriate comments in the pub after work hours. These comments were either derogatory to the contractor or the client, or were construed as indecent comments or behaviour towards members of the public. Although what an employee does outside working-hours should be their own business, it's unfortunately seen as a reflection of their company, more so if they're wearing work clothes bearing the company logo, or driving a company vehicle.

Furthermore, members of the public who are offended by an employee often telephone the client to complain. In some cases they may even have been a potential future client who will now take their business elsewhere.

It's easy for employees to give a company a bad reputation. I've had clients telephone me to complain that the driver of a company vehicle wasn't obeying the road rules, was behaving badly or littering. Sometimes the person wasn't even employed on my project but came from another site.

Clients and their representatives often have dealings with the contractor's workers on site. Although they shouldn't be instructing workers directly they sometimes do, especially when there's a safety concern. All personnel should take the client's safety concerns seriously and respond appropriately, but always politely referring instructions to their Supervisors and managers. Unfortunately I've had employees unhappy with a stranger telling them what they can and can't do, leading to them swearing at the individual who ultimately turns out to be the client. Needless to say they become very irate clients.

Clients appreciate good staff, some even developing an understanding and forming connections with certain of the contractor's staff. They even request these people for their next project, in some cases even specifying that they are allocated to the project as a prerequisite for the contractor to be awarded the contract.

Good staff not only enhances the company's reputation by their actions of delivering a successful, quality project with no incidents, but many of them positively advertise the company to clients, prospective clients, the public and to potential employees as well.

Industrial relations

Clients don't want to do business with a contractor that has industrial relations problems with their workforce or has a reputation of having militant workers because:

- industrial action may disrupt and delay the project
- industrial action and strikes may disrupt the client's operations or their subcontractors' work
- the client's workers may be unduly influenced by the contractor's workers
- industrial action can attract negative press and publicity

Labour harmony is usually one of the client's priorities on a project. It's therefore important that any industrial relations problems are quickly resolved so that they don't escalate and impact on the project or the client's activities.

Reliable equipment

Vehicles and equipment in good repair and newly painted can be a good advert for the company, especially when the company's logo is clearly displayed.

Clients notice the quality of the contractor's equipment and I've often been complimented on our new equipment, even when it wasn't ours but externally hired. Equipment that's poorly maintained, breaks down often, leaks oil, is noisy or smoky creates a poor impression, even when it's externally hired.

Clients prefer equipment in good condition as there is a smaller likelihood of breakdowns which might disrupt and delay the project. There is also less chance of a mechanical failure which could result in an accident or environmental incident.

A word of caution – when selling equipment, ensure that the company's logo and name is removed. I've seen vehicles with the original company name and logo still on them several years after being sold. By now the item is in a poor state of repair and badly rusted. Obviously when the public sees the item they automatically assume it still belongs to the original company. Not only does this portray a poor image of the company, but should the current driver behave or drive badly people will automatically assume they are employed by the company whose name is on the vehicle.

Subcontractors

Subcontractors are seen by the client as being part of the contractor and not separate entities. Good subcontractors assist the contractor in delivering a quality project, safely and on time and are an important part of many projects. The subcontractors that work with a contractor and the manner in which they're managed can influence the contractor's reputation.

Front desk – reception

Many readers may think this doesn't affect them. You may have a small company that cannot afford a front desk, let alone a full time person to answer the telephone. Even more reason to read this section.

The first dealings most people have with a company are through the person that answers the telephone. This person should portray a professional company and needs to be responsive. I've phoned businesses on occasions requiring a quote only for the phone to go unanswered. In many cases I've simply moved on to the next company on the list. In other cases I was able to leave a message, only for no one to call me back. Again I probably took my business elsewhere.

I was fortunate to work for a company that always had pleasant ladies manning the front desk. They greeted people and gave the name of the company when answering the telephone. Calls were transferred efficiently and messages taken and passed on if the person wasn't available.

I'm sure many of you have called a company and asked to talk to an individual. The call is transferred and the person isn't available so you're left holding the telephone with no response from the person that transferred you, or worse still you're left listening to an advert for the company telling you how fast and efficient they are.

Naturally, in many cases all staff act as receptionists for the company, or certainly as a voice for the company. Clients may call Project Managers, Engineers or Supervisors directly. All staff should answer their work telephone in a professional manner. If they are unavailable the caller should be transferred to a message which sounds professional (employees shouldn't use inappropriate voicemail greetings on their company telephones which might sound amusing to their friends but not to others). It's equally important that all staff respond and reply to messages from clients and prospective clients.

Branding

A company's brand is important and major companies go to great expense to sell their brand and protect it. The brand defines how the public and clients perceive the company. If that perception is not accurate you may have to consider working on the brand and changing it.

Case study:

I worked for many years for a company whose name sounded as if we built houses and it was difficult to change the public's perception. When we became known by our acronym rather than the full company name this was no longer a problem.

Another problem was that although the company started off as a civil contractor they became well known as being good commercial builders. Their sign boards were seen on large building projects in the city and much of the company's publicity and advertising was focussed on their building projects. It was difficult for us to sell ourselves to many clients as a serious civil contractor.

Another example of branding is that some car manufacturers have a good name for producing affordable family cars which are popular, yet when they produce a luxury model car, as good as most other luxury cars, it doesn't sell. Why? It's simply about people's perception of the brand.

It's therefore essential to be aware of what your brand is, ensuring it's portrayed correctly. When starting new divisions and new areas of expertise make sure you are able to sell these operations and incorporate them into the brand, but also that they don't detract from and confuse the existing established brand.

Branding often starts with the company's name, and for a company starting out it's often useful to include what the company does. For example 'Smith and Sons' says nothing about what the company does, but 'Smith's Construction', or 'Smith's Electrical Contractors' gives potential clients an indication of what the company does. However be careful of making the name too specific, for example 'Smith's Home Renovations' might not be appropriate in a few years' time when the company is taking on major building projects.

The company name shouldn't be confused with other companies. Also, the name mustn't be too long since it needs to fit onto sign boards and business cards and still remain readable.

Of course there are many experts on branding that can provide useful information, though the information they provide should be appropriate to the industry the company is in and to their client base. After all, it may be pointless branding a construction company and marketing it in bright colours and graphics to an audience at a family weekend event.

Publicity

Unfortunately news reporters are quick to focus on negative incidents and rumours. Accidents, industrial relations incidents, environmental problems and rumours of impropriety quickly turn into news headlines. Sometimes it's not even the contractor's fault but is related to the client or a member of the public. Occasionally the contractor has nothing to do with the problem but the news reporter makes an assumption they are involved. It's very difficult to counter this bad publicity and it's best to:

- endeavour to ensure the company avoids such incidents

- ensure Project Managers report any serious accidents and incidents to senior management so that they are aware of the problem should a reporter call
- have someone from the company available to answer reporters' questions, because if they aren't, reporters will find someone else to talk to who may not be so well informed
- ensure that whoever talks to reporters will be well informed and generally have the same story
- avoid conjecture as to the cause of the problem until an investigation is completed
- make sure that only authorised personnel talk to the reporters
- take rapid action to resolve the problem and to clean up without making the problem bigger
- be seen to investigate problems caused by the contractor
- keep the company's personnel informed of what actually happened, particularly Project Managers who may have to field questions from their clients
- distance the company from actions caused by others

Bad publicity is very difficult to undo and could be a reason for the contractor not being awarded a new project.

Sometimes there are opportunities for good press, so make use of these opportunities by inviting reporters and making sure they have a story to report. This may be for the opening of a newly constructed facility or it could be for a charity event or donation the company has made. Ensure in these instances that the company's logos are clearly displayed and that there's nothing which can detract from the event.

Public relations and neighbours

Most projects impact the neighbouring properties as a result of noise, dust, additional traffic, vehicles parked in the road, diversion of roads and footpaths, and even interruption of services. Many people are fairly accommodating if they are informed beforehand of the construction work.

Some neighbours may be potential future clients and if they become frustrated or inconvenienced by the construction work they won't use the contractor. Furthermore they could tell their friends about problems caused by the contractor, write letters to the newspaper, or complain to the client or town council. Clients have to live with their neighbours long after the contractor has moved on so they certainly don't want a contractor upsetting their neighbours.

Warning the neighbours of the construction work can be done by personally meeting them or by delivering flyers. These could also be used as an opportunity to advertise the services of the company.

Reputation as a good employer

As discussed in Chapter 10 it can be difficult to recruit and retain good staff. This can be made easier if the company has a reputation for being a good employer. To do this it's important the company understands what is important to potential employees, which varies. Some of the things they may look for are a company that:

- treats employees fairly
- rewards people for hard work
- embraces innovation
- offers varied opportunities
- offers personal growth opportunities
- has exciting, large or prestigious projects
- owns new equipment
- employs knowledgeable, experienced people
- provides training and mentoring
- provides travel opportunities
- cares for people and the environment
- is concerned about their staff's welfare
- provides a friendly working environment

One of the best ways of selling the company to potential recruits is by having happy and motivated employees.

Service after the project has been completed

Unfortunately a project doesn't always end when the contractor hands over the keys to the client and moves off site. Most contracts have a warranty period during which the contractor is responsible to repair defects due to their defective workmanship. In fact, in many countries when new structures are build the contractor may even be responsible for defects for several years (five, ten, or longer).

When something goes wrong with a structure the contractor has built the client will inevitably phone the contractor and say that it's the contractor's fault and they must repair it. The contractor needs to respond quickly. Sometimes the response time is dictated by the event – obviously if it's an urgent problem, like a burst water pipe or something which means the client can't use the facility, the response has to be immediate.

Whatever happens in this process, it's important to keep the client informed of the steps being taken to repair the problem since the client usually expects immediate action, even when the problem is minor.

It's usually worth the contractor's time to investigate the problem and ascertain its cause. Many problems highlighted by clients aren't in fact the contractor's fault. Some are design faults, others are general wear-and-tear and some are even caused by the client's poor maintenance regime or operating the facility incorrectly.

Case study:

One of my projects involved the construction of a large concrete bin in which rock was tipped before it was fed into a crusher. A year after the project was commissioned I happened to be talking to another contractor who was still working on the facility. They informed me that the rock kept getting jammed in the bin, so from time to time the operators used explosives to free the rock.

A few weeks later I received a call from the client's designer to report that the concrete in the bin was failing and we needed to repair it.

I sent them a polite letter suggesting that the concrete wasn't designed to resist the blast from explosives and that they should talk to the client and their operations staff.

We never heard anything further regarding the problem.

Unfortunately the problem is often due to the contractor, in which case it needs to be repaired quickly, in a manner which is acceptable to the client and their designers, and with as little disruption to the client's activities as possible. It's usually essential that the repairs are carried out by responsible personnel who will obey the client's rules and carry out the work correctly with little fuss.

Summary

It's imperative that a company has a good reputation and that they are always on their guard to protect it. A poor reputation may result in the company not being awarded projects even when they are the cheapest tenderer.

To create a good reputation means the contractor should:

- avoid making mistakes
- be proactive and assist clients
- be responsive to their clients' demands
- be fair in their dealings with the client
- not charge exorbitant profits
- ensure the work is carried out safely
- produce work of good quality
- be professional
- ensure employees act with honesty and integrity
- make sure that managers visit projects and meets the client regularly
- meet the client's expectations and ensure that these aren't unrealistic
- complete projects on time
- not over-promise and under-deliver
- have committed employees who portray a good image of the company
- have sound industrial relations
- own reliable equipment that's well maintained
- utilise dependable subcontractors that are well managed
- respond to telephone calls in a professional, friendly and helpful manner
- make sure the company's branding is an accurate portrayal of the company
- ensure that negative publicity is dealt with effectively
- make sure that the public and neighbours are kept informed of construction activities affecting them and ensure that these activities create as little disturbance, inconvenience and nuisance as possible

In addition it's useful if companies establish a reputation as a good employer so that they can attract the best personnel.

Conclusion

Managing or running a construction business isn't easy. It's not only about ensuring that projects are successfully completed on time, with no safety incidents, to the correct quality and with a satisfied client, but also about ensuring the company grows in a sustainable way, developing new, viable markets and finding new clients. It's about finding and securing new projects, which means they must be tendered correctly, taking into account all the risks and opportunities of each project. A major error in a tender can destroy a company.

As a company grows it needs to evolve and develop, changing itself, who it deals with, what projects are undertaken, and even how these projects are constructed.

Good staff must be employed, trained, and importantly, retained. They should also share a common passion to deliver quality projects safely and profitably since managers and owners have to rely and trust in them.

The company needs to ensure that revenue is maximised without upsetting clients, and that costs are minimised wherever possible without compromising safety or quality, or at the expense of people or the environment. In all of this, cash flow is extremely important because negative cash flow can destroy a company quicker than an unprofitable project will.

I hope this book has provided some guidance on how to manage a successful and profitable construction company. More important is that this success is sustained and not destroyed by one bad client or one unprofitable project.

Regrettably, I've seen a number of companies fail, putting people out of work, costing suppliers and subcontractors thousands of dollars (sometimes impacting these companies to such an extent that they go insolvent and cause more people to lose their jobs), leaving clients with half-finished projects and banks owed millions. Sometimes the owners of these companies lose everything they own including their house. Many of these failures could have been prevented if the owners and managers had followed some of the suggestions in this book.

As mentioned there are many rules and regulations which govern running a business. These vary between regions and countries and change from time to time, so it's essential that business owners are aware of the latest requirements, and that where necessary they obtain specialist advice.

Glossary

Terminologies vary between different construction industries, countries and even companies. The descriptions below relate more to their meaning within the book and aren't necessarily their official descriptions.

Acceleration – to shorten the schedule, or programme, so the project is completed earlier, or alternatively, to complete more work in the same time period.

Activity – an individual task or event on the schedule.

Allowable – the estimated cost for a particular activity, or task, allowed for in the tender or project budget.

As-built drawings – drawings that are prepared by the contractor to show the position and final dimensions of the structure as constructed.

Back-charges – money charged to the subcontractor for costs the contractor incurred to carry out or rectify the subcontractor's work.

Bond – a form of guarantee issued by a bank or insurance company to insure the client, up to a specified value, should the contractor fail to fulfil their obligations as detailed in the contract.

Change order (variation order) – the written agreement between parties setting out the costs and scope of additional work, or change to the contract.

Claim – a demand from one of the contracting parties for adjustment to the contract.

Civil – construction of concrete structures, roads or railways.

Client – the party who employed and contracted the contractor. The client may be the owner of the facility, the managing contractor, or another contractor. Normally the client is the party that pays the contractor.

Commissioning – the process of testing the equipment and systems installed as part of the construction process.

Construction – the physical work of building or constructing a facility (building, structure, road, dam or factory).

Contour – a line on a map or plan indicating the height of the ground.

Contract – the agreement between the client and contractor.

Contract Administrators – a person who looks after the contractor's project finances, through preparing valuations, claims, cost reports and paying subcontractors.

Contract amendment – specific paperwork used during the construction process should anything change from the original Contract. These changes could be additional work, the omission of work, changes to specifications or the project duration.

Contract document – documents which form the basis of the contract between the parties. They include drawings, terms and conditions and specifications, and set out the requirements for constructing the project.

Contract schedule (contract program/programme) – the schedule which the client has agreed is the official one for the contract, it's used to measure progress, adjudicate any extension of time claims, and if necessary, to quantify the amount of the liquidated damages.

Contractor – a company that constructs or builds a facility or a portion of the facility for a client.

Cost-plus (cost-plus a fee) – when the contractor is reimbursed their actual costs incurred in carrying out the contract, or variation, as well as a mark-up on these costs which is proportional to the costs and is normally expressed as a percentage.

Critical path – a sequence of activities linked together and whose delay will affect the overall project completion.

Day-works – similar to cost-plus. Normally the contractor would specify rates for items of equipment and different tradespeople in the tender document, and these day-works rates are then used to calculate the cost of any additional work which the client may request the contractor to do, which cannot be costed out using the standard tendered unit rates. Often these rates would include a percentage to cover supervision, overheads and profit.

Deliverables – documentation required from the contractor (often required before the project can start)

Demobilisation – the process of moving off site when the project is complete.

Designer – architect or engineer that designs the structures and facilities.

Design and construct contracts – when the contractor is responsible for both the design and the construction of the facility.

Design indemnity insurance – insurance which the designer should have in place in the event that their design is flawed, the structure requires repair or has to be replaced. In many instances the value of this insurance should cover the replacement of the structure.

Drawings (plans) – graphic representation of the structures and facilities.

Due diligence – an audit, or investigation, of the well-being of a company.

Employment contract – the contract between the employer and the employee which defines the conditions of employment.

Escalation – the amount that costs increase with time. These increases are often driven by inflation, wage increases, changes in commodity prices and changes in the value of the local currency.

Estimator – the contractor's person who prepares the tender or estimate.

Exclusions – items which the contractor may have excluded from their tender price.

External hire – equipment that is hired, or rented from another company, or external supplier.

Feasibility study – before deciding to go ahead with a project many clients do an estimate of the project costs to ascertain if the project is viable.

Final account – the final value of the completed work, which includes the original contract value plus all contract amendments and back-charges. The contractor would have a final account with their client, and the contractor should have a final account with each of their subcontractors.

Formwork (shutters) – the forms or structures used to shape and contain the wet concrete used in structures until it has gained sufficient strength to support itself.

Guarantees – a promise or assurance that an obligation will be met.

Insurances – cover for potential losses.

Joint ventures – when two or more contractors enter into an agreement to jointly tender for and contract to a client to construct a facility and in doing so to share resources and risks.

Laydown area – the designated area on a construction site where the contractor can establish their facilities, and store their equipment and materials.

Lead time – the amount of time taken for an item to be delivered to the project. This time includes the time to design, manufacture and transport it to site.

Letter of Intent – a letter issued by the client to the contractor informing them that it is their intention to award them the contract. It's normally treated as an instruction to start the works and the contract document will follow.

Liquidated damages – a specified amount of money which the contractor will pay the client should the contractor fail to meet the agreed contract completion dates.

Litigation – the process of using the court system to resolve a dispute.

Lump sum – many projects are priced as a lump sum contract which usually means the contractor has specified an amount of money to complete the whole project. This amount would include all of the contractor's costs, overheads and profit and is fixed unless the project scope is varied.

Managing contractor – the contractor appointed by the client to manage the project. This could also include a specialist project management company appointed to look after the client's or owner's interests and to manage the design team and the contractor.

Mark-up – the profit margin, although in some cases it may include the profit plus the contractor's overheads.

Materials – all items permanently incorporated by the contractor into the works. This may include concrete, reinforcing, road materials, building products, and including specialist items of equipment.

Mechanical – the construction trades which includes supplying and installing structural steel, mechanical equipment and piping.

Milestone – an important event, such as granting access or a completion date.

Mock-up – a model or small sample built to evaluate details and quality of the final item.

Monthly valuation – an assessment of the work that the contractor has completed during the month which reflects how much the client should pay.

Negotiate – to try and reach an equitable agreement through discussion.

Nominated subcontractor – a subcontractor the client specifies the contractor must use for a particular portion of the work.

Operators – personnel that drive or operate a piece of equipment, a vehicle or machine, but excluding the use of small hand tools.

Over-claim – to claim more money than you're entitled to.

Overhead costs (indirect costs) – overhead project costs are costs the contractor incurs to run the project which cannot be directly related to specific tasks. This includes the provision of management, supervision, site facilities, insurances and bonds. Company overheads are the costs a company incurs which are not directly attributable to a specific project, but are related to running the company and include costs such as Head Office rental, management and various support departments, such as finance and tendering.

Penalties (damages) – many contracts have a provision for the contractor to pay the client an amount of money, or penalty, to cover the client's losses incurred when the contractor fails to finish the project, or sections of the project, by the contracted agreed dates.

Personal protective equipment – equipment issued to personnel for protection at work, this would include safety boots, helmets, gloves, safety glasses and overalls.

Planner (Programmer, or Scheduler) – the person whose specific task is to prepare and update schedules.

Plant – a construction machine (such as an excavator, crane, dozer, loader, and including trucks and vehicles) used on a construction site to perform the work.

Plant and equipment – any item of equipment required to carry out the construction work, but not incorporated into the facility. It includes construction machines, hand tools and formwork.

Post-tender – the period after the contractor has submitted their price (tender or quote) and before the contract or project is awarded to the contractor, and is often the time when the client discusses the contractor's tender and negotiates the final terms, conditions and price of the contract.

Project – any construction work.

Profit (margin or mark-up) – the amount of money the contractor makes after deducting all of their costs from the income or revenue earned.

Project Director – a person who is responsible to manage a number of construction projects and who may have several Project Managers reporting to them.

Project labour agreement – a specific labour agreement established for a project site, which governs the employment conditions (such as hourly rates, working hours, allowances, and project rules), which apply to all workers employed on the project.

Project Manager (site manager, construction manager or site agent) – the person responsible to manage the contractor's work on the construction project.

Punch lists – a list of outstanding items or repairs that must be completed so that the facility complies with the client's requirements.

Quality – the properties of the product supplied to the client, defined by the requirements in the contract document, which may include the visual appearance, as well as, the strength and durability.

Quality control – measures and procedures to ensure the product provided to the client meets the required quality.

Quality plan – the plan drawn up for the contractor to follow to ensure the work meets the required standards and specifications, and to monitor, track and report the procedures implemented by the contractor to ensure the work meets the required quality.

Register – any list or log maintained on the construction site to keep track of items, documents or inspections.

Re-measurable contract – a contract where at the end of the project all of the work completed by the contractor is measured, and providing it meets the specifications, is paid for by the client.

Retention – a portion of money that is owed to the contractor but is withheld by the client, as insurance, until the contractor has fulfilled all their contractual obligations.

Rise-and-fall – a method whereby the contractor is recompensed for the increase in price of a particular item or commodity, calculated from the time when the tender is submitted to when the item's used on the project. As the name implies should the price fall in this period the contractor would pay the difference back to the client.

Scaffolding – temporary structures and platforms to enable workers to reach an elevated work area.

Schedule (often referred to as a programme, program, bar chart or Gantt chart) – a graphic representation of the timetable needed to complete the project, showing the sequencing and duration of the various project tasks and activities.

Scope of works – the work which the contractor is contracted to do. The scope normally takes the form of a written description of the work contained within the contract document.

Self-perform – when the contractor does the work using their own employees rather than using subcontractors.

Shop drawings – drawings produced (normally by the contractor, their suppliers or subcontractors) to show the details of an item they have to fabricate.

Site (project site) – the area where the final construction of the facility takes place.

Site facilities – the contractor's temporary buildings which include offices, toilets, workshops, eating areas, store buildings, and so on.

Slip-form – a construction method where the concrete forms are lifted on a continuous basis while the concrete is poured into them.

Specifications – definitions of the materials, processes and the quality products and systems to be used in the works.

Staff – the contractor's management, supervisory and support personnel who are generally paid a salary and are usually personnel who aren't considered to be workers.

Standards – regulatory codes.

Subcontractor – a contractor employed by a contractor to do a portion of their works. The subcontractor would employ the personnel to do the work.

Superseded drawings – drawings which shouldn't be used for construction since they've been replaced by more up-dated or revised ones.

Supervisor (foreman) – the person who supervisors the contractor's workers or a section of works.

Surety – a form of insurance supplied by a bank or insurance company to ensure that the contractor complies with their contractual obligations.

Survey – to set out the position of structures to be built, or to accurately work out the height and location of existing structures.

Tender (bid, estimate or quote) – a price or quotation to carry out work submitted by the contractor to the client.

Tender covering letter – a letter submitted by the contractor with the tender, this may detail items that the contractor has excluded from their price or assumptions they have made in order to calculate their tender price.

Tender documents (bid documents or quotation) – the documents the client sends to the contractor so they can price the construction of the project. These documents would typically include details of the project, scope of works, specifications and drawings, and sufficient information for the contractor to price the work.

Tender submission (bid submission) – the contractor's response to the tender document, it would include their price, as well as all supporting documentation and any other information, which the client required as part of the tender

Unfixed materials – are materials which have been received by the project but haven't yet been built in, or included in the works.

Union – (trade union) – is a body that represents the worker's rights.

Variation – a change from the original agreed contract.

Warranty – a guarantee that the product will function as it should.

Worker – manual and industrial or trades people who are generally employed on hourly or daily wages, and who physically do the work.

Notes

References

Civitello, Jr. Andrew M & Levy, Sidney M. *Construction Operations Manual of Policies and Procedures: 4th Edition,* McGraw-Hill

Dykstra, Alison. *Construction Project Management: A Complete Introduction,* Kirshner Publishing Company, INC.

Ganaway, Nick B. *Construction Business Management: What every Construction Contractor, Builder and Subcontractor needs to Know,* Butterworth-Heinemann an imprint of Elsevier

Gerstel, David. Running a Successful Construction Company. The Taunton Press

Halpin, Daniel W & Senior, Bolivar A. *Construction Management: 4th Edition,* Hamilton Printing

Jackson, Barbara J. *Construction Management Jumpstart: 2nd Edition,* Sybex an Imprint of Wiley

Levy, Sydney M. *Project Management in Construction. 6th Edition:* McGraw-Hill

Mincks, William R & Johnston, Hal. *Construction Jobsite Management: 2nd Edition,* Thomson Delmar Learning

Mubarak, Saleh A. *Construction Project Scheduling and Control. 2nd Edition:* John Wiley and Son Schexnayder, Clifford J & Mayo, Richard E. *Construction Management Fundamentals,* McGraw-Hill

Netscher, Paul. *Successful Construction Project Management: The Practical Guide.* Panet Publications

Stevens, Matt. *The Construction MBA, Practical approaches to construction contracting.* McGraw-Hill

Stone, Michael C. *Markup & Profit – A Contractor's guide.* Craftsman book Company

Walker, Anthony. *Project Management in Construction: 5th Edition,* Blackwell Publishing

Also by Paul Netscher

Successful Construction Project Management: The Practical Guide

- **Are you looking for an easy to read practical guide to project manage a construction project?**
- **Are you tired of project management books that focus on the theoretical and seem to have little relevance to your project?**
- **Do your Project Managers need a little extra help and guidance?**

Successful Construction Project Management is a valuable companion to Building a Successful Construction Company. Written by the same author the book is aimed at both aspiring Project Managers as well as those who are more experienced. This easy to read book avoids being overly 'text book' in style and is filled with practical everyday examples incorporating 28 years of construction experience gained on over 120 projects in 6 countries. It's written by a construction professional for construction professionals.

Those who have read the book comment:

'I highly recommend this book be read by all newly qualified construction project managers as well as those more experienced.' *(Customer 1 on Amazon uk)*
'Easy Reading' 'I have been looking for a book like this. A great study guide for the novice project manager and the more experienced project managers.' *(Customer 2 Amazon uk)*
'It is a very easy to use book with guidelines that are referenced intelligently with case studies.' *(Customer 3 Amazon)*
'Well worth a read' *(Customer 4 Amazon)*
'Having read a few project management texts over the years it is great to find one that doesn't just want to run you through the stock standard theory of the process.' *(Dellas Lynch)*

The book includes chapters on planning the project, starting it, project schedule, managing the project, closing it out, quality, safety, people, materials, equipment, financial and contractual matters. Each chapter is divided into multiple subsections allowing the reader to easily pick issues relevant to their current project. Many topics aren't included in other construction management books despite their importance.

Made in the USA
Middletown, DE
11 February 2021

33515930R00152